GENERAL PROPERTIES OF MATTER

General Properties of Matter

B. BROWN, B.Sc., Ph.D., F.Inst.P

Senior Lecturer in Physics
University of Salford

Springer Science+Business Media, LLC

Published in the U.S.A. by
PLENUM PRESS
a division of
PLENUM PUBLISHING CORPORATION
227 West 17th Street, New York, N.Y. 10011

First published by
Butterworth & Co. (Publishers) Ltd.

ISBN 978-1-4899-6237-9   ISBN 978-1-4899-6501-1 (eBook)
DOI 10.1007/978-1-4899-6501-1

Suggested U.D.C. number: 531/532
Suggested additional number 539
Library of Congress Catalogue Card Number 75-84056

PREFACE

In recent years university courses in physics have been constantly reviewed and revised in order to provide a balanced course which includes the essentials of classical physics, as well as the latest discoveries and theories of modern physics. In many colleges this revision has meant that the time allocated to the subject of 'properties of matter' has been somewhat reduced from that formerly available. It is perhaps also unfortunate that in the schools the subject is often neglected in favour of the more glamorous topics such as nuclear physics, with the result that the student embarking on a university physics course often has a very sketching knowledge of properties of matter. It might be added that his knowledge of nuclear physics is also frequently very sketchy! This present book has been written in order to provide a concise introductory textbook on properties of matter suitable for use by first year university physics students, bearing in mind that such students may have only a limited acquaintance with the subject.

In many universities the current trend is to provide the first year student with a course of lectures on properties of matter together with a separate course of lectures on waves and vibrations. This book deals with the topics usually included under the first heading and it does not include any account of waves and vibrations which are adequately dealt with in a companion volume. The contents of a course of lectures, on properties of matter, vary from one university to another. In a few universities, courses in properties of matter may well include the topics of kinetic theory and relativity. More often, however, these topics are regarded as being important enough to be dealt with in separate courses of lectures. The author subscribes to this view and hence they are not dealt with in the present volume, as other available books deal with them in a much more complete manner than would have been possible in a book of this size.

Since this book is intended for first year students no mathematical knowledge is required of the student other than that normally dealt with in current 'A' level courses. Only in the treatment of Hydrodynamics does it prove more convenient to use vector notation and this chapter may conveniently be left until the student has acquired a sufficient knowledge of vector methods.

B.B.

CONTENTS

1

PARTICLE DYNAMICS

1.1 RECTILINEAR MOTION

CONSIDER the motion of a particle along a straight line OS, the point O being fixed. Let the positions of the particle at times t_1 and t_2 be P_1 and P_2 respectively, *Figure 1.1*. If $OP_1 = s_1$ and $OP_2 = s_2$ the average velocity of the particle during the time interval considered is $s_2 - s_1/t_2 - t_1$. If this quantity is the same for all time intervals the particle velocity is said to be uniform.

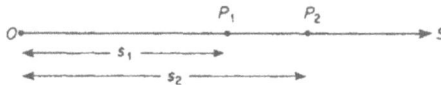

Figure 1.1. Rectilinear motion of a particle

Consider a particle moving along the line OS so that at times t and $t + \delta t$ it is distance s and $s + \delta s$ from O respectively. Then, during the time interval δt the velocity is $\delta s/\delta t$. As $\delta t \to 0$ the limiting value of $\delta s/\delta t$ is the velocity of the particle at time t. Thus

$$v = \lim_{\delta t \to 0} \frac{\delta s}{\delta t} = \frac{\mathrm{d}s}{\mathrm{d}t} = \dot{s} \qquad \dots (1.1)$$

1.2 ACCELERATION

If v_1 and v_2 are the velocities of a moving particle at time t_1 and t_2 respectively, the average acceleration, i.e. rate of change of velocity during the time interval, is $v_2 - v_1/t_2 - t_1$. If this quantity is the same for all time intervals the particle acceleration is said to be uniform.

Consider a particle moving so that at times t and $t + \delta t$ its velocities are v and $v + \delta v$ respectively. The average value of acceleration during the time δt is $\delta v/\delta t$. As $\delta t \to 0$ the limiting value of $\delta v/\delta t$ is the acceleration at time t.

1

Thus
$$a = \lim_{\delta t \to 0} \frac{\delta v}{\delta t} = \frac{dv}{dt} = \dot{v} = \ddot{s}$$
.... (1.2)

1.3 UNIFORMLY ACCELERATED MOTION IN ONE DIRECTION

If a particle is moving along a straight line with uniform acceleration then the equation of motion is

$$\ddot{s} = a$$
.... (1.3)

i.e.
$$\frac{d^2s}{dt^2} = a$$

Integrating,
$$\frac{ds}{dt} = at + c_1$$
.... (1.4)

If $\frac{ds}{dt} = u$ when $t = 0$, $c_1 = u$, and

$$v = \frac{ds}{dt} = at + u$$
.... (1.5)

Integrating again, $s = ut + \frac{at^2}{2} + c_2$
.... (1.6)

If $s = 0$ at $t = 0$ then $c_2 = 0$, so that

$$s = ut + \frac{at^2}{2}$$
.... (1.7)

Alternatively, writing equation (1.3) as

$$v \frac{dv}{ds} = a$$
.... (1.8)

Integrating,
$$\frac{v^2}{2} = as + c_3$$
.... (1.9)

Since $s = 0$ when $t = 0$ and $v = u$

$$c_3 = \frac{u^2}{2}$$
.... (1.10)

Thus
$$v^2 = u^2 + 2as$$
.... (1.11)

It will be seen that equations (1.5), (1.7) and (1.11) are the standard equations for uniformly accelerated motion in a straight line.

1.4 UNIFORMLY ACCELERATED MOTION IN ONE DIRECTION UNDER A RESISTANCE PROPORTIONAL TO VELOCITY

In this case the equation of motion is

$$\frac{d^2s}{dt^2} = a - kv = \frac{dv}{dt} \qquad \dots (1.12)$$

i.e.
$$\frac{dv}{a - kv} = dt \qquad \dots (1.13)$$

Integrating, $\frac{-1}{k} \log (a - kv) = t + c_1 \qquad \dots (1.14)$

If $v = 0$ when $t = 0$, $c_1 = -\frac{1}{k} \log a$, so that

$$\log \frac{a - kv}{a} = -kt$$

i.e.
$$v = \frac{a}{k}(1 - e^{-kt}) \qquad \dots (1.15)$$

Hence
$$\frac{k}{a} ds = (1 - e^{-kt}) dt \qquad \dots (1.16)$$

Integrating,
$$\frac{k}{a} s = t + \frac{1}{k} e^{-kt} + c_2 \qquad \dots (1.17)$$

If $s = 0$ when $t = 0$, $c_2 = -\frac{1}{k}$, so that

$$s = \frac{a}{k^2}(e^{-kt} - 1 + kt) \qquad \dots (1.18)$$

From this equation it can be seen that as t becomes very large e^{-kt} becomes vanishingly small and v approaches its maximum possible value under the given conditions. This value, i.e. a/k is called the terminal velocity.

1.5 UNIFORMLY ACCELERATED MOTION IN ONE DIRECTION UNDER A RESISTANCE PROPORTIONAL TO THE SQUARE OF THE VELOCITY

In this case the equation of motion is

$$\frac{d^2s}{dt^2} = a - kv^2 = a\left(1 - \frac{kv^2}{a}\right) \qquad \dots (1.19)$$

3

When $v^2 = \dfrac{a}{k}, \dfrac{d^2s}{dt^2} = 0$ and hence the maximum value of v, i.e., its

terminal velocity is $\sqrt{(a/k)}$.

Let $\sqrt{(a/k)} = V$, then substituting in equation (1.19)

$$\frac{dv}{dt} = a\left(1 - \frac{v^2}{V^2}\right) \qquad \qquad \dots(1.20)$$

Hence $\qquad \dfrac{at}{V^2} = \displaystyle\int \dfrac{dv}{V^2 - v^2} = \dfrac{1}{2V}\log\dfrac{V + v}{V - v} + c_1 \qquad \dots(1.21)$

If $v = 0$ at $t = 0$, $c_1 = 0$, so that

$$\frac{V + v}{V - v} = e^{2at/V^2} \qquad \qquad \dots(1.22)$$

$$v = V.\frac{e^{at/v} - e^{-(at/v)}}{e^{at/v} + e^{-(at/v)}} = V.\tanh\left(\frac{at}{V}\right) \qquad \dots(1.23)$$

From equation (1.23)

$$\int ds = V\int \tanh\frac{at}{V}\,dt$$

$$s = \frac{V^2}{a}\log\cosh\frac{at}{V} + c_2 \qquad \qquad \dots(1.24)$$

If $s = 0$ when $t = 0$, $c_2 = 0$ and hence

$$s = \frac{V^2}{a}\log\cosh\frac{at}{V} \qquad \qquad \dots(1.25)$$

1.5.1 Example

A particle moves in a medium in which the resistance is proportional to the square of the particle velocity. It is projected vertically upwards against gravity with velocity v. Find the time taken for the particle to return.

Considering upward motion, the equation of motion is

$$\frac{du}{dt} = -g - ku^2 \qquad \qquad \dots(1.26)$$

4

Now if V is the terminal velocity

$$\frac{du}{dt} = -\frac{g}{V^2}(V^2 + u^2) \qquad \dots (1.27)$$

$$\frac{du}{V^2 + u^2} = -\frac{g}{V^2}\,dt$$

$$\tan^{-1}\frac{u}{V} = -\frac{gt}{V} + A \qquad \dots (1.28)$$

Now from equation (1.27)

$$u\frac{du}{ds} = -\frac{g}{V^2}(V^2 + u^2) \qquad \dots (1.29)$$

$$\log(V^2 + u^2) = -\frac{2gs}{V^2} + B \qquad \dots (1.30)$$

If $s = 0$ when initial velocity is v, then

$$B = \log(V^2 + v^2) \text{ and } A = \tan^{-1}\frac{v}{V}$$

Thus $\qquad s = \dfrac{V^2}{2g}\log\dfrac{V^2 + v^2}{V^2 + u^2} \qquad \dots (1.31)$

When the particle comes to rest

$$s = \frac{V^2}{2g}\log\frac{V^2 + v^2}{V^2} \text{ and } t = \frac{V}{g}\tan^{-1}\frac{v}{V} \qquad \dots (1.32)$$

In the subsequent descent, from equation (1.25)

$$s = \frac{V^2}{g}\log\cosh\frac{gt}{V}$$

$$t = \frac{V}{g}\cosh^{-1} e^{gs/V^2} \qquad \dots (1.33)$$

In this case

$$s = \frac{V^2}{2g}\log\frac{V^2 + v^2}{V^2}$$

$$t = \frac{V}{g}\cosh^{-1}\left(\frac{V^2 + v^2}{V^2}\right)^{\frac{1}{2}} \qquad \dots (1.34)$$

5

B

Hence, the total time elapsing before the particle returns is, from equations (1.32) and (1.34)

$$\frac{V}{g} \left[\tan^{-1} \frac{v}{V} + \cosh^{-1} \left(\frac{V^2 + v^2}{V^2} \right)^{\frac{1}{2}} \right] \qquad \dots (1.35)$$

1.6 THE LAWS OF MOTION

In the seventeenth century Newton postulated three laws of motion and these form the basis of dynamics. These laws can be stated as follows:

1. Every body continues in its state of rest or uniform motion in a straight line, except in so far is it is compelled to change that state by forces impressed upon it.

2. The change of motion is proportional to the impressed force and takes place in the straight line in which the force acts.

3. An action is always opposed by an equal reaction.

From the first law, force is defined as that which changes or tends to change the motion of a body. Newton defines the 'motion' of a body as: 'The quantity of motion of a body is the measure of it arising from its velocity and the quantity of matter conjointly', the motion referred to is momentum. According to the second law, if m is the mass of a particle moving with a velocity v, its momentum is mv. If a force F acting on the particle produces an acceleration a, then

$$F \propto \frac{d}{dt}(mv)$$

Assuming the mass to be constant

$$F \propto m \frac{dv}{dt}$$
$$\propto ma$$

If the unit of force is that which produces unit acceleration when acting on unit mass, then

$$F = ma = m \frac{d^2s}{dt^2} \qquad \dots (1.36)$$

In the c.g.s. system of units the unit of force is the dyne, while in the f.p.s. system it is a poundal. In the SI system the unit of force is the Newton.

6

The weight W of a body is given by

$$W = mg$$

where m is its mass and g is the acceleration due to gravity. So that a one pound weight is equal to g poundals, a one gramme weight is equal to g dynes, and a one kilogram weight is equal to g newtons, the actual value of g being appropriate to the system of units.

1.7 IMPULSE

If a particle is acted upon by a constant force F so that its velocity changes from u to v in a time t during which period a distance s is traversed then

$$v - u = at$$

and
$$v^2 - u^2 = 2as \qquad \dots (1.37)$$

Hence

$$mv - mu = mat = Ft$$

and
$$\tfrac{1}{2}mv^2 - \tfrac{1}{2}mu^2 = mas = Fs \qquad \dots (1.38)$$

Ft is called the impulse and from equation (1.37), it is evident that the impulse is equal to the change of momentum. Thus the impulse of a force is defined as the change of linear momentum produced by it. Equation (1.38) shows that the change in the kinetic energy of a particle is equal to the work done by the force. The work done by a force is not always transferred into kinetic energy, sometimes the force is opposed by another force and some of the energy is expended in overcoming this force. Under these circumstances part of the work done raises the potential energy and hence the total energy of a particle is partly kinetic and partly potential. It is measured relative to some arbitrary zero of velocity and position.

1.8 CONSERVATIVE SYSTEMS OF FORCES

A conservative system of forces is defined as follows: when the forces acting on a system of bodies are of such a nature that the algebraic work done in performing any series of displacements, whereby the original configuration of the system is regained, is zero, then those forces constitute a conservative system of forces.

Consider (*i*) the work done in raising a mass m against gravity through a vertical distance h, and (*ii*) the work done in pushing a body against a frictional force F through a distance s. In case (*i*) the work done against gravity is mgh. If the mass is allowed to fall to its original position then the available work is mgh. Thus the total work done on the mass is zero. In case (*ii*) the work done is Fs and if the body is to be brought back to its original position the same amount of work must be done so that the total expenditure of work is $2Fs$. In processes similar to case (*i*) work can be recovered from the system by making the system of bodies perform mechanical work as they return to their original positions, and such systems constitute a conservative system of forces. In processes similar to case (*ii*) no recovery of expended work is possible and such systems are non-conservative.

Magnetostatic, electrostatic and gravitational forces are conservative, while forces due to friction or the resistance of a medium in which a body is moving are non-conservative.

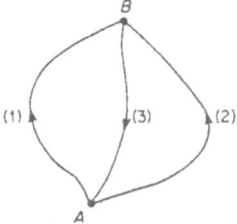

Figure 1.2. A conservative system of forces

The work done by a system of conservative forces in moving a system of bodies from a configuration A to a second one B, is independent of the path from A to B. This is easily shown by reference to *Figure 1.2*. Let the work done in passing from A to B along paths (1) and (2) be W_1 and W_2 respectively. Let W_3 be the work done in passing from B to A along path (3). Now since the forces constitute a conservative system

$$W_1 + W_3 = 0 \text{ and } W_2 + W_3 = 0$$

Hence
$$W_1 = W_2$$

1.9 POTENTIAL ENERGY OF A SYSTEM
The potential energy of a system is defined as follows: If any configuration A is taken as standard, the work done when a system of bodies, under the action of conservative forces, passes from the

configuration A to a configuration B, is called the potential energy of the system in the configuration B. Thus the work stored in the system when it has taken up the configuration B is measured by the potential energy of that configuration.

If the potential energies appropriate to the configurations A and B are W_1 and W_2 respectively, then the work done against conservative forces in moving the system of bodies from configuration A to B is $W_2 - W_1$. This can be shown as follows: W_1 represents the work done when the system passes from some standard configuration to the configuration A, and W_2 is a measure of the work done when the system passes from the same standard configuration to the configuration B; but W_2 is the work from the standard configuration to A plus the work from A to B, hence the work from A to B is $W_2 - W_1$.

1.10 TOTAL ENERGY

Equation (1.38) showed that the change in the kinetic energy of a particle is equal to the work done by the force acting on it. If a system is considered where the initial position and velocity of a particle of mass m are s_1 and $\dot{s}_1$ respectively at time t_1 and the final position and velocity are s_2 and $\dot{s}_2$ at time t_2, then

$$\tfrac{1}{2}m(\dot{s}_2^2 - \dot{s}_1^2) = \int_{s_1}^{s_2} F \, ds \qquad \dots (1.39)$$

For a conservative system of forces $F = -(dV/ds)$ where V is the potential energy. Thus

$$\int_{s_1}^{s_2} F \, ds = -\int_{s_1}^{s_2} \frac{dV}{ds} \, ds = V_1 - V_2 \qquad \dots (1.40)$$

$$\tfrac{1}{2}m\dot{s}_2^2 + V_2 = \tfrac{1}{2}m\dot{s}_1^2 + V_1 \qquad \dots (1.41)$$

Equation (1.41) shows that the sum of the kinetic and potential energies is constant and this sum is termed the total energy. It measures the amount of work the body can perform against external agencies in passing from its actual state regarding velocity and position to rest in some standard position. The above proof is done for only one variable, s, but is, of course, true in general.

1.11 TWO-DIMENSIONAL MOTION

So far it has been assumed that the position of a particle varied with time in one direction only. An extension of the arguments used is

9

necessary when the motion of a particle in two dimensions is considered. The equations involved are considerably simplified if polar coordinates are used instead of cartesian coordinates. In general the particle will have both radial and transverse velocities and accelerations. Radial velocity refers to the rate of increase of the radius vector while transverse velocity refers to the velocity of the particle in a direction at right angles to the radius vector.

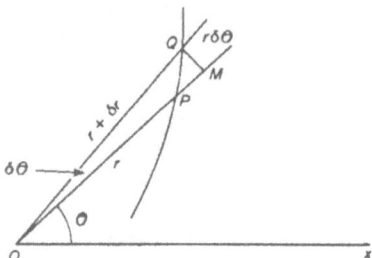

Figure 1.3. Radial and transverse velocities

With reference to *Figure 1.3*, let P be the position of a particle at time t and Q its position at time $t + \delta t$. Let $XOP = \theta$, $XOQ = \theta + \delta\theta$, $OP = r$ and $OQ = r + \delta r$. QM is perpendicular to OP. Let u be the radial velocity and v the transverse velocity, hence

$$u = \lim \frac{MP}{\delta t} = \lim \frac{(r + \delta r)\cos \delta\theta - r}{\delta t}$$

$$= \lim \frac{\delta r}{\delta t} = \frac{dr}{dt} = \dot{r} \qquad \qquad \dots (1.42)$$

$$v = \lim \frac{MQ}{\delta t} = \lim \frac{r\delta\theta}{\delta t}$$

$$= r \frac{d\theta}{dt} = r\dot{\theta} \qquad \qquad \dots (1.43)$$

With reference to *Figure 1.4*, the radial and transverse velocities at P, u and v, become $u + \delta u$ and $v + \delta v$ respectively at Q. Thus the radial acceleration along OP is

10

$$\lim\frac{(u + \delta u)\cos\delta\theta - (v + \delta v)\sin\delta\theta - u}{\delta t} = \lim\frac{u + \delta u - v\delta\theta - u}{\delta t}$$

$$= \frac{du}{dt} - v\frac{d\theta}{dt} = \frac{d^2r}{dt^2} - r\left(\frac{d\theta}{dt}\right)^2$$

$$= \ddot{r} - r\dot\theta^2 \qquad \dots (1.44)$$

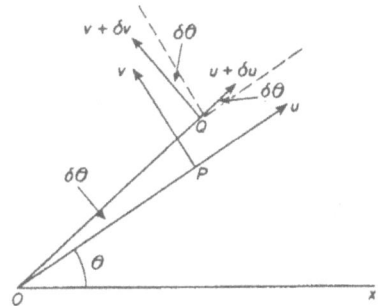

Figure 1.4. Radial and transverse accelerations

The transverse acceleration perpendicular to OP is

$$\lim\frac{(v + \delta v)\cos\delta\theta + (u + \delta u)\sin\delta\theta - v}{\delta t} = \frac{dv}{dt} + u\frac{d\theta}{dt}$$

$$= \frac{d}{dt}\left(r\frac{d\theta}{dt}\right) + \frac{dr}{dt}\cdot\frac{d\theta}{dt} = r\frac{d^2\theta}{dt^2} + 2\frac{dr}{dt}\cdot\frac{d\theta}{dt}$$

$$= r\ddot\theta + 2\dot r\dot\theta = \frac{1}{r}\frac{d}{dt}(r^2\dot\theta) \dots (1.45)$$

For motion in a circle r = constant. If the speed is constant and equal to $rd\theta/dt$ then the radial acceleration is $-r(d\theta/dt)^2$ in a direction away from the centre, while the transverse acceleration is zero.

1.12 CENTRAL ORBITS

When a particle is moving under an acceleration which is always directed towards a fixed point its path is called a central orbit. At any instant, the particle is moving in a plane containing the radius vector from the fixed point to the particle, and the tangent to the orbit at the point considered. From the definition, all central orbits have zero transverse acceleration, hence from equation (1.45)

11

$$\frac{1}{r}\frac{d}{dt}(r^2\dot\theta) = 0$$

i.e.
$$r^2\dot\theta = \text{constant} = A \qquad \dots (1.46)$$

If the acceleration at a point (r, θ) is equal to a and directed towards the centre, then equation (1.44) for radial acceleration becomes

$$\ddot r - r\dot\theta^2 = -a \qquad \dots (1.47)$$

Let $\dfrac{1}{r} = u$

Then
$$\frac{dr}{dt} = \frac{d}{dt}\left(\frac{1}{u}\right) = \frac{d}{d\theta}\left(\frac{1}{u}\right)\frac{d\theta}{dt} = \frac{1}{u^2}\cdot\frac{du}{d\theta}\cdot\frac{d\theta}{dt}$$

Also $A = r^2\dot\theta = \dfrac{1}{u^2}\dot\theta$

Hence $\dfrac{dr}{dt} = -A\dfrac{du}{d\theta}$

Now
$$\frac{d^2 r}{dt^2} = \frac{d}{dt}\left(-A\frac{du}{d\theta}\right) = -A\frac{d}{d\theta}\left(\frac{du}{d\theta}\right)\frac{d\theta}{dt} = -A\frac{d^2 u}{d\theta^2}\cdot\frac{d\theta}{dt}$$

$$= -A^2 u^2 \frac{d^2 u}{d\theta^2}$$

Substituting for $\dfrac{d^2 r}{dt^2}$ and $\dfrac{d\theta}{dt}$ in equation (1.47)

$$-a = -A^2 u^2 \frac{d^2 u}{d\theta^2} - \frac{1}{u}A^2 u^4 \qquad \dots (1.48)$$

or
$$\frac{a}{A^2 u^2} = \frac{d^2 u}{d\theta^2} + u \qquad \dots (1.49)$$

This is the general differential equation to any orbit.

At any point in the orbit the areal velocity is defined as

$$\lim\frac{\text{area described by radius vector}}{\text{time taken}}$$

From *Figure 1.3* it is evident that area *POQ* is described in time δt, hence the areal velocity is

$$\lim \text{area} \frac{POQ}{\delta t} = \lim \frac{\frac{1}{2}r(r + \delta r)\, sm\delta\theta}{\delta t} = \tfrac{1}{2}r^2\dot\theta = \frac{A}{2} \qquad \dots (1.50)$$

Thus the radius vector describes equal areas in equal times.

1.13 MOTION UNDER AN INVERSE SQUARE LAW OF ATTRACTION

Provided the law governing the relationship between the central acceleration and the distance of the particle from the centre is known, then, after substituting for a in equation (1.49), the equation can be solved to give the family of curves to which the orbit belongs. The selection of a particular member of the family of curves is determined by the initial conditions. In the case where the acceleration is inversely proportional to the square of the distance, then

$$a = \frac{\omega}{r^2} = \omega u^2$$

and the equation of motion becomes

$$\omega u^2 = A^2 u^2 \left[\frac{d^2 u}{d\theta^2} + u \right] \qquad \dots (1.51)$$

$$\frac{d^2 u}{d\theta^2} + u - \frac{\omega}{A^2} = 0 \qquad \dots (1.50)$$

Put $\qquad\qquad u - \dfrac{\omega}{A^2} = \overline{U}$

Then $\qquad \dfrac{d\overline{U}}{d\theta} = \dfrac{du}{d\theta} \quad$ and $\quad \dfrac{d^2\overline{U}}{d\theta^2} = \dfrac{d^2 u}{d\theta^2}$

On substitution $\qquad \dfrac{d^2\overline{U}}{d\theta^2} + \overline{U} = 0 \qquad \dots (1.51)$

or $\qquad\qquad \dfrac{1}{2}\dfrac{d}{d\overline{U}}\left(\dfrac{d\overline{U}}{d\theta}\right)^2 + \overline{U} = 0$

$$\int d\left(\frac{d\overline{U}}{d\theta}\right)^2 = -\int 2\overline{U}\,d\overline{U}$$

Hence $\qquad \left(\dfrac{d\overline{U}}{d\theta}\right)^2 = -\dfrac{2\overline{U}^2}{2} + C \qquad \dots (1.52)$

When $\qquad \overline{U} = 0 \quad$ Let $\dfrac{d\overline{U}}{d\theta} = D$

13

B*

Thus
$$\left(\frac{d\bar{U}}{d\theta}\right)^2 = D^2 - \bar{U}^2$$

and
$$\frac{d\bar{U}}{d\theta} = \sqrt{(D^2 - \bar{U}^2)} \qquad \dots (1.53)$$

Rearranging
$$\int \frac{d\bar{U}}{\sqrt{(D^2 - \bar{U}^2)}} = \int d\theta$$

Hence
$$\sin^{-1}\frac{\bar{U}}{D} = \theta + \text{constant} = \theta + \varepsilon \qquad \dots (1.54)$$

$$\therefore \ \bar{U} = D \sin(\theta + \varepsilon) \qquad \dots (1.55)$$

$$\therefore \ u - \frac{\omega}{A^2} = D \sin(\theta + \varepsilon)$$

$$\frac{1}{r} - \frac{\omega}{A^2} = D \sin(\theta + \varepsilon)$$

Let $\dfrac{\omega}{A^2} = \dfrac{1}{l}$

Then
$$\frac{l}{r} = 1 + Dl \sin(\theta + \varepsilon)$$

$$\therefore \ \frac{l}{r} = 1 + \mathscr{E} \cos(\theta - \gamma) \qquad \dots (1.56)$$

where $\varepsilon = Dl$ and γ is another constant. Equation (1.56) is the polar equation of a conic (see proof below). Hence the central orbit of a particle moving under the inverse square law is a conic with one focus at the centre of acceleration. The eccentricity $\mathscr{E}$ of the conic is given by

$$\mathscr{E} = Dl = \frac{DA^2}{\omega} \qquad \dots (1.57)$$

and the semi-latus rectum by

$$l = \frac{A^2}{\omega} \qquad \dots (1.58)$$

To identify any particular case it is necessary to know whether the eccentricity is greater than, equal to, or less than unity and this depends on the initial conditions determining M and γ.

14

For example, consider a body projected with velocity v from a point distance a from the centre of force in a direction making angle α with the radius vector. This is illustrated in *Figure 1.5.*

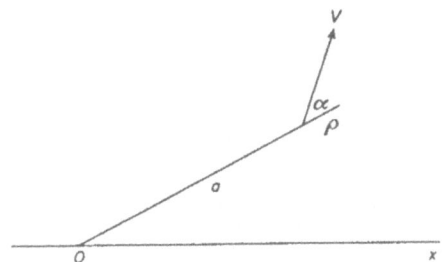

Figure 1.5. Determination of eccentricity

Now initially, when $\theta = 0, r = a, \dfrac{dr}{dt} = v \cos \alpha$ and $\dfrac{d\theta}{dt} = \dfrac{v \sin \alpha}{a}$

Thus
$$A = r^2 \dot{\theta} = av \sin \sin \alpha$$

Substituting in equation (1.56)

$$\frac{a^2 v^2 \sin^2 \alpha}{a\omega} = 1 + \mathscr{E} \cos (\theta - \gamma) \qquad \dots (1.57)$$

Differentiating equation (1.56)

$$\frac{l}{r^2}\frac{dr}{d\theta} = \mathscr{E} \sin (\theta - \gamma)$$

Thus initially

$$\frac{a^2 v^2 \sin^2 \alpha}{a^2 \omega} \cdot \frac{av \cos \alpha}{v \sin \alpha} = \mathscr{E} \sin (\theta - \gamma) \qquad \dots (1.58)$$

$$\frac{v^2 a \sin \alpha \cos \alpha}{\omega} = \mathscr{E} \sin (\theta - \gamma) \qquad \dots (1.59)$$

Squaring equations (1.57) and (1.59)

$$\mathscr{E}^2 \cos^2 (\theta - \gamma) = \frac{a^2 v^4 \sin^4 \alpha}{\omega^2} - \frac{2av^2 \sin^2 \alpha}{\omega} + 1$$

and $\qquad \mathscr{E}^2 \sin^2 (\theta - \gamma) = \dfrac{v^4 a^2 \sin^2 \alpha \cos^2 \alpha}{\omega^2}$

Thus, adding

$$\mathscr{E}^2 = 1 + \frac{v^4 a^2 \sin^2 \alpha}{\omega^2}(\sin^2 \alpha + \cos^2 \alpha) - \frac{2av^2 \sin^2 \alpha}{\omega}$$

$$= 1 + \frac{v^4 a^2 \sin^2 \alpha}{\omega^2} - \frac{2av^2 \sin^2 \alpha}{\omega} \qquad \qquad \dots (1.60)$$

Thus, the orbit is an ellipse, parabola or hyperbola according to whether v^2 is less than, equal to, or greater than $2\omega/a$; this condition is independent of the direction of projection. The result is of importance in connection with the movements of heavenly bodies.

1.14 POLAR EQUATION OF A CONIC

With reference to *Figure 1.6*, let S be the focus of a conic, SZ the normal from S to the directrix ZA, P a point on the curve and M the

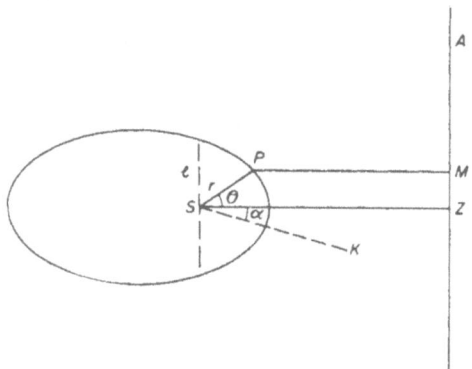

Figure 1.6. Polar equation of a conic

projection of P in the directrix. Let $SP = r$ and θ be the angle between SP and SZ. If $\mathscr{E}$ is the eccentricity of the conic, then

$$SP = \mathscr{E} . PM$$

If l is the semi-latus rectum

$$l = \mathscr{E} . SZ = \mathscr{E}(r \cos \theta + PM)$$
$$= \mathscr{E}r \cos \theta + r$$

Hence $\dfrac{l}{r} = 1 + \mathscr{E} \cos \theta$

16

If the initial line to which the orientation of the radius vector SP is referred is SK, making an angle γ with SZ, then $\angle PSZ = (\theta - \gamma)$ and the polar equation of the conic becomes

$$\frac{l}{r} = 1 + \mathscr{E}\cos(\theta - \gamma) \qquad \dots (1.61)$$

1.15 PERIODIC TIME IN THE CASE OF AN ELLIPTIC ORBIT

This is the time required for a radius vector to sweep out an area equal to that of the ellipse. If a and b are the semi-axes of the ellipse, its area is πab. It has already been shown that the areal velocity, i.e. the area swept out in unit time, is equal to $A/2$, see equation (1.50). Hence the period T is given by

$$T = \frac{\pi ab}{A/2} \qquad \dots (1.62)$$

Now the semi-latus rectum of the ellipse is related to the semi-axes by the equation

$$l = \frac{b^2}{a}$$

Also, by definition $\qquad l = \dfrac{A^2}{\omega}$

$$A^2 = \frac{\omega b^2}{a} \qquad \dots (1.63)$$

Substituting for A in equation (1.62) we find

$$T^2 = \frac{4\pi^2 a^3}{\omega} \qquad \dots (1.64)$$

Hence the square of the period is proportional to the cube of the major axis of the ellipse. Since the expression is independent of b, the semi-minor axis of the ellipse, the periodic time is the same for all elliptic orbits having equal major semi-axes.

1.16 KEPLER'S LAWS

Kepler stated his famous laws in 1618. These three laws dealt with the shape of a planet's orbit in its motion round the sun, the relationship between the orbital speed and the corresponding distance from the sun, and the connection between the size of the planetary

17

orbit and the year of the planet respectively. Kepler put forward these laws as a result of astronomical observations and did not assume a special law of gravitational attraction. The laws may be stated as follows:

1. Each planet moves in an ellipse with the sun at one focus.
2. The line joining a planet to the sun traces out equal areas in equal times, i.e. the areal velocity of the radius vector is constant.
3. The square of the time for the completion of one circuit of the orbit (the year of the planet) is proportional to the cube of the major axis of the orbit.

In view of the results derived in the preceding sections this means that the planets are under the action of central accelerations directed towards the sun, and that this central acceleration varies with distance according to the inverse square law. The correspondence between Kepler's empirical laws and the results derived on the basis of the Newtonian law of gravitation indicates that over planetary distances, gravitation follows the Newtonian law. In fact, sufficient experimental evidence has been accumulated to show that the Newtonian law of gravitational attraction is much more general and breaks down only at distances of less than molecular magnitude.

1.16.1 Examples

(1) At any point P outside the Earth and at a distance s from its centre, the acceleration a due to gravity is inversely proportional to s^2. If R is the radius of the Earth and g is the value of a at its surface, express a in terms of R, g and s, Find the speed in miles per hour with which an artificial satellite would describe a circle of radius 4500 miles, concentric with the Earth, and the time it would take to describe the circle once. Take R as 3960 miles and g as 79,000 miles/hour/h.

Let $a = k/s^2$ where k is a constant.

At the Earth's surface $a = g = k/R^2$, $\qquad k = gR^2$

and hence $\qquad\qquad\qquad a = \dfrac{gR^2}{s^2}$

If a satellite at P travelling with velocity v executes a circular orbit about the Earth, it must have a component of acceleration v^2/s towards 0, the centre of the Earth, produced by the gravitational attraction of the Earth upon it

18

$$\frac{mv^2}{s} = \frac{mgR^2}{s^2}$$

$$v^2 = \frac{gR^2}{s}$$

$$v = R\sqrt{\frac{g}{s}}$$

$$= 16{,}600 \text{ m.p.h.}$$

The time taken to circle the Earth once

$$= \frac{2\pi s}{v} = 2\pi \cdot \frac{4500}{16{,}600}$$

$$= 1 \text{ h } 42 \text{ min}$$

(2) A particle of mass m is moving under the action of a force directed towards a fixed point which varies as a/r^3, where a is a constant and r is the distance of the particle from the fixed point. If the angular momentum of the particle about the fixed point is J, find the equation of the orbit.

The transverse acceleration is zero.

Hence

$$\frac{1}{r}\frac{d}{dt}(r^2\dot{\theta}) = 0$$

From the radial acceleration $= \ddot{r} - r\dot{\theta}^2$

$$m\ddot{r} - mr\dot{\theta}^2 = -\frac{a}{r^3} \qquad \dots (1.65)$$

Now

$$J = mr^2\dot{\theta} \qquad \dots (1.66)$$

Substituting in equation (1.65)

$$m\ddot{r} = \left(\frac{J^2}{m} - a\right)\frac{1}{r^3} \qquad \dots (1.67)$$

Now

$$\ddot{r} = \frac{d}{dt}(\dot{r}) = \frac{d\theta}{dt}\cdot\frac{d}{d\theta}\left(\frac{dr}{d\theta}\cdot\frac{d\theta}{dt}\right)$$

$$= \dot{\theta}\frac{d}{d\theta}\left(\dot{\theta}\frac{dt}{d\theta}\right)$$

$$= \frac{J}{mr^2}\frac{d}{d\theta}\left(\frac{J}{mr^2}\frac{dr}{d\theta}\right)$$

$$= \frac{J^2}{m^2}\cdot\frac{1}{r^2}\frac{d}{d\theta}\left(\frac{1}{r^2}\frac{dr}{d\theta}\right)$$

19

Let $u = \dfrac{1}{r}$

$$\therefore \quad \frac{dr}{d\theta} = -\frac{1}{u^2}\frac{du}{d\theta}$$

Hence
$$\ddot{r} = \frac{J^2}{m^2}\cdot u^2 \frac{d}{d\theta}\left(-\frac{du}{d\theta}\right) \qquad \ldots (1.68)$$

Substituting in equation (1.67)

$$\frac{J^2 u^2}{m}\frac{d^2 u}{d\theta^2} = u^3\left(a - \frac{J^2}{m}\right) \qquad \ldots (1.69)$$

Hence
$$\frac{d^2 u}{d\theta^2} = -u\left(1 - \frac{am}{J^2}\right)$$

The solution of which is

$$u = A\sin(n\theta - \varepsilon)$$

Where
$$n = \sqrt{\left(1 - \frac{am}{J^2}\right)}$$

Hence the equation of the orbit is

$$\frac{1}{r} = A\sin(n\theta - \varepsilon) \qquad \ldots (1.70)$$

1.17 MOTION OF A PROJECTILE ASSUMING NO FRICTION

Projectile motions under gravity are more conveniently treated in cartesian coordinates and the velocities and accelerations along the x and y axes are treated separately. This distinction is possible because a force cannot affect motion in a direction at right angles to its own line of action.

Consider a particle projected from the origin of coordinates with velocity v at an angle θ to the horizontal axis. The motion along OX is not accelerated and the equation of motion in this direction is

$$\frac{dx}{dt} = v\cos\theta \qquad \ldots (1.71)$$

Since $x = 0$ when $t = 0$

$$x = vt\cos\theta \qquad \ldots (1.72)$$

In the vertical direction

20

$$\frac{d^2y}{dt^2} = -g$$

also, when $t = 0$

$$\frac{dy}{dt} = v \sin \theta$$

$$\therefore \frac{dy}{dt} = v \sin \theta - gt \qquad \dots (1.73)$$

When $t = 0$, $y = 0$, so

$$\therefore y = vt \sin \theta - \frac{gt^2}{g} \qquad \dots (1.74)$$

Eliminating t from equations (1.72) and (1.74), the equation to the trajectory is

$$y = x \tan \theta - \frac{gx^2}{2v^2 \cos^2\theta} \qquad \dots (1.75)$$

This represents a parabola and the particle describes a parabolic orbit.

To find the range on a plane inclined at angle α to the horizontal, the point of intersection of the trajectory and the plane must be found.

Thus from the equations

$$y = x \tan \theta - \frac{gx^2}{2v^2 \cos^2 \theta}$$

and $\qquad\qquad\qquad y = x \tan \alpha$

It can be shown that

$$x = \frac{2v^2 \cos^2 \theta}{g}(\tan \theta - \tan \alpha) \qquad \dots (1.76)$$

and $\qquad y = \frac{2v^2 \cos^2 \theta \tan \alpha}{g}(\tan \theta - \tan \alpha) \qquad \dots (1.77)$

The range R is given by

$$R^2 = x^2 + y^2 \qquad \dots (1.78)$$

On a horizontal plane $\theta = 0$ and the horizontal range is therefore

$$R = x = \frac{2v^2 \sin \theta \cos \theta}{g} = \frac{v^2 \sin^2 2\theta}{g}$$

To determine the coordinates of the highest point reached by the particle in its path, as well as the time taken, apply the condition

$$\frac{dy}{dt} = 0$$

Thus from equation (1.73), at this point

$$t = \frac{v \sin \theta}{g} \qquad \dots (1.79)$$

From equation (1.74)

$$y = \frac{v^2 \sin^2 \theta}{g} - \frac{v^2 \sin^2 \theta}{2g} = \frac{v^2 \sin^2 \theta}{2g}$$

Also

$$x = \frac{R}{2} = \frac{v^2 \sin^2 2\theta}{2g}$$

After time t has elapsed the particle has velocity components dy/dt and dx/dt. Hence the resultant velocity is given by

$$v_1^2 = \left(\frac{dy}{dt}\right)^2 + \left(\frac{dx}{dt}\right)^2$$

Its direction makes an angle β with the x axis, so that

$$\tan \beta = \frac{dy}{dx} = \frac{dy}{dt} \frac{dt}{dx}$$

Hence from equations (1.71) and (1.73)

$$\tan \beta = \frac{v \sin \theta - gt}{v \cos \theta} = \tan \theta - \frac{gt}{v \cos \theta}$$

and $\qquad v_1^2 = v^2 - 2vgt\sin \theta + g^2t^2 \qquad \dots (1.81)$

1.18 THE MOTION OF A PROJECTILE SUBJECTED TO A RESISTANCE PROPORTIONAL TO ITS VELOCITY

In this case, the equations of motion along the x and y axes become

$$\frac{d^2x}{dt^2} = -k\frac{dx}{dt} \qquad \dots (1.82)$$

$$\frac{d^2y}{dt^2} = -k\frac{dy}{dt} - g \qquad \ldots . (1.83)$$

Let
$$\frac{dx}{dt} = v'$$

Thus
$$\frac{dv'}{dt} = -kv'$$

and
$$\log v' = -kt + c$$

When $t = 0$, $v' = v\cos\theta$, hence

$$c = \log(v\cos\theta)$$

$$\therefore \frac{dx}{dt} = e^{-kt}v\cos\theta \qquad \ldots . (1.84)$$

and
$$x = -\frac{v\cos\theta}{k}e^{-kt} + c$$

Since $x = 0$ when $t = 0$,

$$c = \frac{v\cos\theta}{k}$$

$$\therefore x = \frac{v\cos\theta}{k}(1 - e^{-kt}) \qquad \ldots . (1.85)$$

Let $\dfrac{dy}{dt} = v_2$

Thus
$$\frac{dv_2}{dt} = -kv_2 - g$$

Integrating, $\log(kv_2 + g) = -kt + c$

$$= -kt + \log(kv\sin\theta + g)$$

Since when $t = 0$, $v_2 = v\sin\theta$

Then
$$kv_2 + g = (kv\sin\theta + g)e^{-kt} \qquad \ldots . (1.86)$$

Integrating,
$$ky + gt = -\left(\frac{kv\sin\theta + g}{k}.e^{-kt}\right) + c$$

when $y = 0$, $t = 0$, thus

$$c = \frac{kv\sin\theta + g}{k}$$

23

Hence $$ky + gt = \frac{kv\sin\theta + g}{k}(1 - e^{-kt}) \qquad \dots (1.87)$$

1.18.1 Example

A particle is projected from the top of a cliff 128 ft high with a velocity of 140 ft/s at an elevation of $\tan^{-1} 4/3$. Find the distance from the foot of the cliff at which it strikes the sea and show that its direction of motion then makes an angle of approximately 59°45′ with the horizontal.

Let the distance from the foot of the cliff be x and let the angle made by its direction of motion with the horizontal be α.

Considering vertical motion, if t is the time of flight, then applying equation (1.74)

$$-128 = vt \sin\theta - 16t^2$$

i.e. $$-128 = 140 \cdot \tfrac{4}{5}t - 16t^2$$

From which $$t = 8 \text{ s}$$

Considering horizontal motion, from equation (1.72)

$$x = vt \cos\theta$$
$$= 140 \cdot \tfrac{3}{5} \cdot 8$$
$$= 672 \text{ ft}$$

Horizontal component of velocity at sea level $= v \cos\theta = 84$ ft/s. Vertical component of velocity is given by

$$v' = v\sin\theta - gt = -144 \text{ ft/s}$$

Hence $$\tan\alpha = \tfrac{144}{84} \simeq 1 \cdot 7143$$
$$\therefore \ \alpha = 59°45′$$

2

MOTION OF A RIGID BODY

2.1 MOTION OF A SYSTEM OF PARTICLES

LET $(x_1, y_1), (x_2, y_2)$, etc. be the coordinates of a system of particles of masses m_1, m_2, etc. Let the particles be subject to external forces whose components parallel to the axes are X_1, Y_1, X_2, Y_2, etc. Let the particles also be subject to internal motions and reactions due to the mutual actions of the particles upon one another of which the components on m_1, m_2 etc. are X'_1, Y'_1, X'_2, Y'_2, etc.

The equations of motion for the separate particles are

$$\left. \begin{array}{l} m_1\ddot{x}_1 = X_1 + X'_1, m_1\ddot{y}_1 = Y_1 + Y'_1 \\ m_2\ddot{x}_2 = X_2 + X'_2, m_2\ddot{y}_2 = Y_2 + Y'_2 \end{array} \right\} \quad \ldots (2.1)$$

Adding, there results,

$$\left. \begin{array}{l} \sum m\ddot{x} = \sum X + \sum X' \\ \sum m\ddot{y} = \sum Y + \sum Y' \end{array} \right\} \quad \ldots (2.2)$$

By the law of reaction, Newton's third law, the internal actions and reactions are equal and opposite in pairs, so that the sums of the resolved parts in any direction must vanish. Hence

$$\sum X' = 0 \text{ and } \sum Y' = 0$$

Thus equations (2.2) become

$$\sum m\ddot{x} = \sum X \text{ and } \sum m\ddot{y} = \sum Y \quad \ldots (2.3)$$

i.e. the rate of change of the linear momentum of the whole system in any prescribed direction is equal to the sum of the resolved parts of the external forces in that direction.

If there is a direction in which the sum of the resolved parts of the external forces is zero, the rate of change of linear momentum in that direction must also be zero.

For example, if $\sum X = 0$ then $\sum m\ddot{x} = 0$.

25

By integration therefore

$$\sum m\dot{x} = \text{const.} \qquad \ldots (2.4)$$

i.e. the linear momentum in that direction is constant. This is the principle of conservation of linear momentum.

Considering equations (2.1), multiply each x equation by the corresponding y and each y equation by the corresponding x. There results

$$m_1\ddot{x}_1 y_1 = X_1 y_1 + X'_1 y_1, m_1\ddot{y}_1 x_1 = Y_1 x_1 + Y'_1 x_1$$
$$m_2\ddot{x}_2 y_2 = X_2 y_2 + X'_2 y_2, m_2\ddot{y}_2 x_2 = Y_2 x_2 + Y'_2 x_2 \qquad \ldots (2.5)$$

Subtraction gives

$$m_1(x_1\ddot{y}_1 - y_1\ddot{x}_1) = x_1 Y_1 - y_1 X_2 + x_1 Y'_1 - y_1 X'_1$$
$$m_2(x_2\ddot{y}_2 - y_2\ddot{x}_2) = x_2 Y_2 - y_2 X_2 + x_2 Y'_2 - y_2 X'_2 \text{ etc.} \qquad \ldots (2.6)$$

Addition of those equations gives

$$\sum m(x\ddot{y} - y\ddot{x}) = \sum(xY - yX) + \sum(xY' - yX') \qquad \ldots (2.7)$$

but the moment about the origin of a vector (*Figure 2.1*), whose components are X, Y, located at the point $(x, t\, y)$ is

$$xY - yX$$

Thus $\sum(xY' - yX')$ is the sum of moments about the origin of the internal actions and reactions. Since these are equal and opposite in pairs

$$\sum(xY' - yX') = 0 \qquad \ldots (2.8)$$
$$\therefore \qquad \sum m(x\ddot{y} - y\ddot{x}) = \sum(xY - yX) \qquad \ldots (2.9)$$

Alternatively equation (2.9) may be written

$$\frac{d}{dx}\sum m(x\dot{y} - y\dot{x}) = \sum(xY - yX) \qquad \ldots (2.10)$$

i.e. the rate of change of moment of momentum of the system about any fixed origin is equal to the sum of the moments of the external forces about that origin.

If the sum of the moments of the external forces about any fixed origin or axis is zero then the moment of momentum of the system about that axis is constant.

26

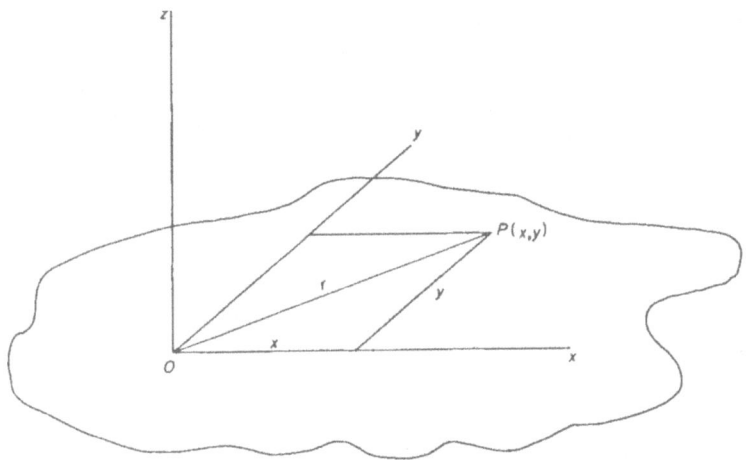

Figure 2.1. Perpendicular axes theorem for a laminar body

Moment of momentum is also termed angular momentum and the above sentence is a statement of the principle of conservation of angular momentum.

2.2 INDEPENDENCE OF TRANSLATION AND ROTATION
Consider a system of particles whose total mass is M and let $\bar{x}, \bar{y}$ be the coordinates of the centre of gravity G. The equations determining the position of the centre of gravity are

$$M\bar{x} = \sum mx, \quad M\bar{y} = \sum my \qquad \dots (2.11)$$

Differentiation of these equations gives

$$M\dot{\bar{x}} = \sum m\dot{x}, \quad M\dot{\bar{y}} = \sum m\dot{y} \qquad \dots (2.12)$$

i.e. the linear momentum of the system is the same as that of a particle whose mass is the whole mass, moving with the velocity of the centre of gravity. Differentiation of equations (2.12) gives

$$M\ddot{\bar{x}} = \sum m\ddot{x}, \quad M\ddot{\bar{y}} = \sum m\ddot{y} \qquad \dots (2.13)$$

Thus from equation (2.3)

$$M\ddot{\bar{x}} = \sum X, \quad M\ddot{\bar{y}} = \sum Y \qquad \dots (2.14)$$

i.e. the motion of the centre of gravity G is the same as if all the mass

27

were collected into a particle at G and all the external forces were moved parallel to themselves to act at G.

Therefore, if a system is not acted upon by external forces, its centre of gravity is at rest, or moving with constant velocity, since integration of $M\ddot{\bar{x}} = 0$ and $m\ddot{\bar{y}} = 0$, gives $M\dot{\bar{x}} = $ const. and $M\dot{\bar{y}} = $ const.

If in equation (2.9) we put

$$x = \bar{x} + x' \quad y = \bar{y} + y'$$

where x', y' denote the coordinates of the particle of mass m relative to the centre of gravity G then, from equation (2.11)

$$\sum mx' = \sum my' = 0$$

Thus

$$\sum m\ddot{x}' = \sum m\ddot{y}' = 0$$

Substituting in equation (2.9)

$$\sum m\{(\bar{x} + x')(\ddot{\bar{y}} + \ddot{y}') - (\bar{y} + y')(\ddot{\bar{x}} + \ddot{x}')\}$$
$$= \sum \{(\bar{x} + x')Y - (\bar{y} + y')X\} \qquad \ldots (2.15)$$

Multiplying out and eliminating $\sum mx\ddot{y}' = \bar{x}\sum m\ddot{y}' = 0$, etc., gives

$$M(\bar{x}\ddot{\bar{y}} - \bar{y}\ddot{\bar{x}}) + \sum m(x'\ddot{y}' - y'\ddot{x}')$$
$$= \bar{x}\sum Y - \bar{y}\sum X + \sum(x'Y - y'X) \qquad \ldots (2.16)$$

From equation (2.14)

$$M\ddot{\bar{x}} = \sum X \text{ and } M\ddot{\bar{y}} = \sum Y$$

So equation (2.16) becomes

$$\sum m(x'\ddot{y}' - y'\ddot{x}') = \sum(x'Y - y'X) \qquad \ldots (2.17)$$

This equation shows that the rate of change of angular momentum about the centre of gravity is equal to the sum of the moments of the external forces about the centre of gravity.

Thus it is seen that the motion of the centre of gravity and of the system relative to the centre of gravity are independent of each other and this is the principle of the independence of translation and rotation.

2.3 CONSERVATION OF ENERGY

If the equations (2.1) are multiplied by the corresponding velocity component there results

$$m_1\ddot{x}_1\dot{x}_1 = X_1\dot{x}_1 + X_1'\dot{x}_1, m_1\ddot{y}_1\dot{y} = Y_1\dot{y}_1 + Y_1'\dot{y}_1 \text{ etc.} \qquad \ldots (2.18)$$

A'ddition of these equations gives

$$\sum m(\dot{x}\ddot{x} + \dot{y}\ddot{y}) = \sum\{(X + X')\dot{x} + (Y + Y')\dot{y}\}$$

which can be written,

$$\frac{d}{dt}\{\tfrac{1}{2}\sum m(\dot{x}^2 + \dot{y}^2)\} = \sum\{(X + X')\dot{x} + (Y + Y')\dot{y}\} \quad\ldots(2.19)$$

In this equation the left-hand term is the rate of increase of kinetic energy of the system; the right-hand term is the rate at which the external and internal forces are doing work. The increase in kinetic energy in any time is therefore equal to the work done.

In a case where the potential energy depends on the configuration of the system and where a change of potential energy due to a change of configuration is independent of the way in which the change is made, then the work done will be equal to the loss of potential energy. Therefore the sum of the kinetic energy and potential energy remains constant.

2.4 RIGID BODIES

A rigid body consists of a number of particles bound together by cohesive forces and internal mutual attractions which are always equal and opposite. The results obtained above are therefore also true for a rigid body, thus the motion of the centre of gravity of a rigid body is independent of its size and shape and depends only on its mass and the resultant of the external forces acting on it. The size and shape of the rigid body affect its rotational motion relative to the centre of gravity, since the angular momentum and the moment of the forces acting depend on the size and shape.

The principles established previously may be restated as applied to a rigid body as follows:

(1) The rate of change of linear momentum of a rigid body in any direction is equal to the sum of the components of the external forces resolved in that direction.

(2) The rate of change of angular momentum of a rigid body about any fixed axis is equal to the sum of the moments of the external forces about that axis.

(3) The increase in the kinetic energy of a rigid body in any time is equal to the work done by the external forces in that time. The principles of the independence of translation and rotation are also valid for a rigid body.

2.5 MOMENT OF INERTIA

If the mass of every element of a body or particle of a system is multiplied by the square of its distance from an axis, then the sum

29

of the products is called the 'moment of inertia' of the body or system about that axis. If m denotes the mass of an element or particle and r denotes its distance from the axis, then I, the moment of inertia is given by

$$I = \sum mr^2 \qquad \dots (2.20)$$

If M denotes the whole mass of the body or system and k is a line, of length such that Mk^2 is equal to the moment of inertia about an axis, then k is called the radius of gyration of the body or system about that axis.

2.6 THEOREM OF PARALLEL AXES

This states that the moment of inertia of a body about any axis is equal to its moment of inertia about a parallel axis through its centre of gravity together with the product of the whole mass and the square of the distance between the axes.

Let the given axis be the z axis. Let the coordinates of a particle of mass m be (x,y,z), the coordinates of the centre of gravity G be $(\bar{x},\bar{y},\bar{z})$, and let $x = \bar{x} + x'$, $y = \bar{y} + y'$ and $z = \bar{z} + z'$.

The moment of inertia of the body about the z axis is

$$\sum m(x^2 + y^2) = \sum m\{(\bar{x} + x')^2 + (\bar{y} + y')^2\}$$
$$= \sum m(x'^2 + y'^2) + (\bar{x}^2 + \bar{y}^2)\sum m + 2\bar{x}\sum mx' + 2\bar{y}\sum my'$$

$$\dots (2.21)$$

but, from equation (2.11)

$$\sum mx' = 0 \text{ and } \sum my' = 0$$

and therefore the moment of inertia of the body about the z axis is

$$\sum m(x'^2 + y'^2) + (\bar{x}^2 + \bar{y}^2)\sum m \qquad \dots (2.22)$$

In this equation the-first sum is the moment of inertia about an axis through G parallel to the z axis and the remaining terms are the product of the whole mass and the square of the distance between the axes.

2.7 PERPENDICULAR AXES THEOREM

The moment of inertia of a plane lamina, about an axis perpendicular to its plane, is equal to the sum of the moments of inertia about any two perpendicular axes in the plane which intersect on the first axis.

Let the plane of the lamina be in the XY plane and the perpendicular axis be the Z axis. Then with reference to *Figure 2.1*, let the moments of inertia of the laminar body about the three mutually perpendicular axes, Ox, Oy, Oz, be I_x, I_y, I_z respectively. Let a particle at the point $P(x,y)$ have mass m. Then

$$I_x + I_y = \sum m(x^2 + y^2) = \sum mr^2 = I_z \qquad \ldots (2.23)$$

2.8 CALCULATION OF MOMENTS OF INERTIA

In many cases the value of the moment of inertia may be found by simple integration, i.e. if dm represents an infinitesimal part of the whole mass and is situated at a distance r from the axis of rotation, then

$$I = \int r^2 \, \mathrm{d}m \qquad \ldots (2.24)$$

(The limits of the integral are chosen to cover the whole of the body.)

(1) Moment of inertia of a uniform thin rod about an axis at one end perpendicular to its length.

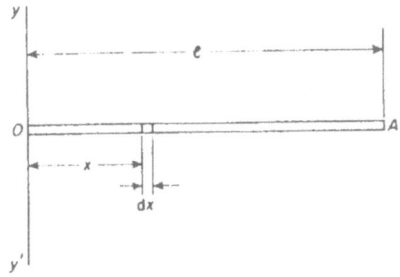

Figure 2.2. Moment of inertia of a uniform thin rod about an axis at one end perpendicular to its length

Consider a rod, length l, having a uniform linear distribution of mass m. The moment of inertia of an element of length dx situated at a distance x from the axis (see Figure 2.2), is mdx. x^2, and the total moment of inertia is given by

$$I = \int_0^l mx^2 \, \mathrm{d}x = \frac{Ml^2}{3} \qquad \ldots (2.25)$$

since M, the mass of the rod is equal to ml.

(2) Moment of inertia of a uniform thin rod about an axis normal to its length and passing through the centre.

If one half of the rod is considered the problem is identical with case (1) and

$$I_{\frac{1}{2}} = \int_0^{l/2} mx^2 \, dx$$

Hence, for the whole rod

$$I = 2 \int_0^{l/2} mx^2 \, dx$$
$$= \frac{Ml^2}{12} \qquad \qquad \ldots (2.26)$$

(3) Moment of inertia of a uniform circular lamina about an axis through its centre and normal to its plane.

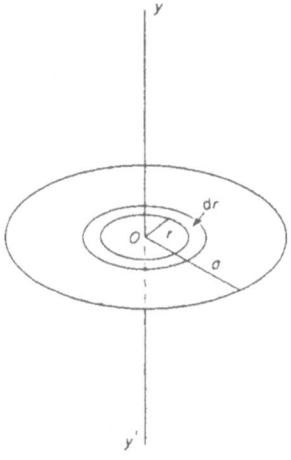

Figure 2.3. Moment of inertia of a uniform circular lamina about an axis through its centre and normal to its plane

With reference to *Figure 2.3*, divide the disk into thin circular rings. Consider one of the rings of radius r and width dr. Let the mass per unit area of the disk be m.

Thus, the moment of inertia of this ring is

$$2\pi r \, dr \, . \, mr^2$$

The total moment of inertia is given by

$$I = \int_0^a 2\pi mr^3 \, dr$$
$$= \frac{Ma^2}{2} \qquad \qquad \ldots (2.27)$$

where M is the mass of the disk $= \pi a^2 m$.

In the case of an annular disk of inner and outer radii b and a respectively

$$I = \int_b^a 2\pi r^3 m \, dr = \frac{2\pi m(a^4 - b^4)}{4}$$

Now M, the mass of the disk $= \pi m(a^2 - b^2)$

therefore $$I = \frac{M}{2}(a^2 + b^2)$$ .... (2.28)

(4) Moment of inertia of a rectangular lamina about an axis through its centre and perpendicular to its plane.

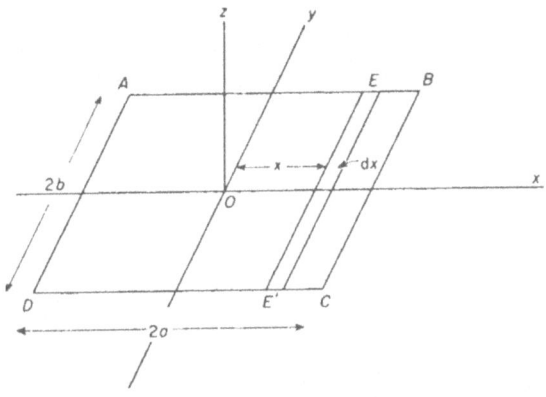

Figure 2.4. *Moment of inertia of a rectangular lamina about an axis through its centre and perpendicular to its plane*

With reference to *Figure* 2.4, let O be the centre of mass of the lamina and let axes Ox, Oy be chosen so that they are in the plane of the lamina and parallel to the sides of length $2a$ and $2b$ respectively. Let the mass per unit area be m and let EE' be an element of the lamina parallel to Oy at a distance x from it, and of width dx. The mass of this element is $2bdxm$ and its moment of inertia about Oy is given by $2bdxm \cdot x^2$.

Hence the total moment of inertia about Oy is given by

$$I_y = 2\int_0^a 2bdxmx^2 = \frac{4mba^3}{3}$$

33

but the mass of the lamina $M = 4abm$ so that

$$I_y = \frac{Ma^2}{3} \qquad \qquad \ldots(2.29)$$

Similarly,
$$I_x = \frac{Mb^2}{3} \qquad \qquad \ldots(2.30)$$

Now by the theorem of perpendicular axes

$$I_z = I_x + I_y$$

$$\therefore \qquad I_z = \frac{M}{3}(a^2 + b^2) \qquad \qquad \ldots(2.31)$$

(5) Moment of inertia of a circular or annular disk about a diameter.

It has previously been shown (3) that the moment of inertia of a circular lamina about an axis through its centre and normal to its plane is $Ma^2/2$. If this axis is Oz and if Ox and Oy are two axes at right angles to one another and to Oz,

$$I_x = I_y = \frac{I_z}{2} = \frac{Ma^2}{4} \qquad \qquad \ldots(2.32)$$

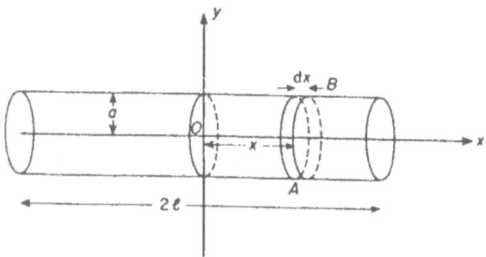

Figure 2.5. *Moment of inertia of a uniform cylinder about an axis through its centre and at right angles to its length*

Similarly, in the case of an annular disk

$$I_x = I_y = \frac{I_z}{2} = \frac{M}{4}(a^2 + b^2) \qquad \qquad \ldots(2.33)$$

(6) Moment of inertia of a uniform cylinder about an axis through its centre and at right angles to its length.

34

With reference to *Figure 2.5*, let the length of the cylinder be $2l$ and its radius a. Let O be the centre of mass of the cylinder, Ox the axis of revolution, Oy the axis at right angles to Ox about which the moment of inertia is to be calculated and let the density of the cylinder material be ρ.

Let the circular element AB be at distance x from O and be of thickness dx. The moment of inertia of this disk about Ox is then given by

$$\pi a^2 dx \rho \cdot \frac{a^2}{2}$$

Thus the moment of inertia about a diameter is

$$\frac{\pi a^4 \rho dx}{4}$$

and its moment of inertia about Oy is

$$\pi a^2 dx \rho x^2 + \frac{\pi a^4 \rho dx}{4}$$

Therefore the total moment of inertia of the cylinder about Oy is

$$I = 2 \left\{ \int_0^l \pi a^2 \rho x^2 dx + \frac{\pi a^4 \rho dx}{4} \right\}$$

$$= 2\pi a^2 \rho \left[\frac{l^3}{3} + \frac{a^2 l}{4} \right]$$

$$= M \left(\frac{l^2}{3} + \frac{a^2}{4} \right) \qquad \ldots (2.34)$$

since the mass of the cylinder $M = \pi a^2 2l\rho$

(7) Moment of inertia of a uniform sphere about a diameter.

With reference to *Figure 2.6*, let O be the centre of the sphere of radius a of which the material density is ρ. Consider an element of the sphere formed by dividing the sphere by planes normal to Ox, the axis about which the moment of inertia is required. Let the element be of width dx and at a distance x from O. If the radius of this element is y, its mass m is $\pi y^2 dx \rho$ and its moment of inertia about Ox is $my^2/2$. Hence, the moment of inertia of the whole sphere about Ox is given by

$$I = 2 \int_0^a \pi y^2 \rho \frac{y^2}{2} dx$$

Since
$$a^2 = x^2 + y^2$$

$$I = \pi\rho \int_0^a (a^2 - x^2)^2 \, dx$$

$$= \pi\rho \left[a^5 - \frac{2a^5}{3} + \frac{a^5}{5} \right]$$

$$= \frac{8}{15} \pi\rho a^5$$

Since the mass of the sphere $M = \frac{4}{3} \pi a^3 \rho$

$$I = \frac{2}{5} M a^2$$

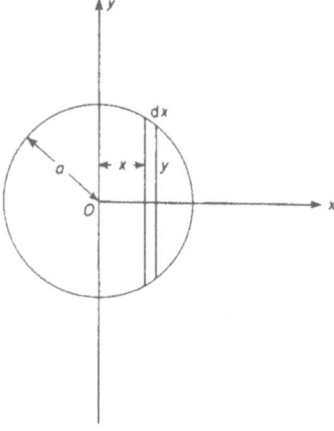

Figure 2.6. Moment of inertia of a uniform solid sphere about a diameter

By the theorem of parallel axes, the moment of inertia of the solid sphere about a tangent I_T is given by

$$I_T = I + M a^2$$

$$= \frac{7Ma^2}{5} \qquad\qquad \ldots\ldots (2.35)$$

(8) Moment of inertia of a uniform bar of rectangular cross-section about an axis through its centre of mass.

With reference to *Figure 2.7*, let O be the centre of mass of the rectangular bar. Let the three axes Ox, Oy, Oz, be drawn through O parallel to the edges of the bar. Let the edges of the bar be of length

36

2a, 2b and 2c respectively, and let the material of the bar have density ρ.

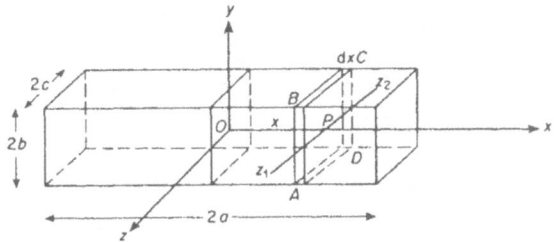

Figure 2.7. Moment of inertia of a uniform rectangular bar about the axis Oz

Let P be the centre of mass of an element $ABCD$ of the bar, thickness dx and at distance x from O. The moment of inertia of this element of mass $4bc\rho dx$, about the axis $Z_1 P Z_2$ parallel to Oz, is $4bc\rho dx. \, b^2/3$.

Thus, by the theorem of parallel axes, the moment of inertia of the element about the axis Oz is

$$4bc\rho dx\left(\frac{b^2}{3} + x^2\right)$$

Hence the moment of inertia of the whole bar about Oz is given by

$$I = 2\int_0^a 4bc\rho dx\left(\frac{b^2}{3} + x^2\right)$$

$$= 8bc\rho\left[\frac{ab^2}{3} + \frac{a^3}{3}\right]$$

$$= \frac{M}{3}(a^2 + b^2) \qquad\qquad \dots (2.36)$$

since the mass of the bar $M = 8abc\rho$.

If part of a body is removed, the moment of inertia of the body about any axis is reduced in value by the moment of inertia of the part removed. Thus, the moment of inertia of a body containing a cavity is found by taking the difference between the moment of inertia of the whole body and of the part which would fill the cavity.

2.9 ROUTH'S RULE

It has previously been stated that the radius of gyration of a body about an axis is a quantity such that $I = Mk^2$, where I is the moment

37

c

of inertia of the body about that axis and M is the mass of the body. The values of many radii of gyration may easily be determined according to Routh's rule, which applies to linear, plane and solid homogeneous bodies. These are classified as α rectangular, β circular or elliptical, or γ spherical, spheroidal or ellipsoidal.

Routh's rule states that the radius of gyration about an axis of symmetry passing through the mass centre of the body is given by

$$k^2 = \frac{\text{sum of squares of perpendicular semi-axes}}{3, 4, \text{ or } 5}$$

where the denominator is 3, 4, or 5 according to whether the body falls into the α, β, or γ classification.

2.10 EXAMPLES

(1) A uniform body consists of a solid hemisphere of radius a with a right circular cone whose base radius and height are also a, attached to it so that the two circular faces coincide. Show that the moment of inertia about an axis through the circumference of the common base, perpendicular to those bases, is equal to $41\,Ma^2/30$ where M is the total mass of the body.

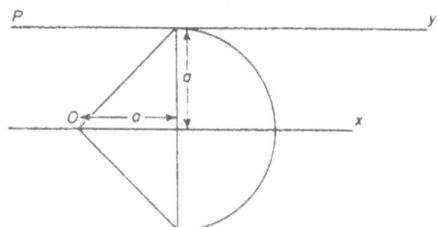

Figure 2.8. Moment of inertia problem

With reference to *Figure 2.8*, let the density of the material be ρ.

	Cone	Hemisphere	Whole
Mass	$\frac{1}{3}\pi a^3 \rho = M_1$	$\frac{2}{3}\pi a^3 \rho = M_2$	$\pi a^3 \rho = M$
Formula for the moment of inertia about the axis	$\dfrac{3M_1 a^2}{10}$	$\dfrac{2M_2 a^2}{5}$	—
Moments of inertia about the axis Ox	$\dfrac{3}{10}\dfrac{\pi a^3 \rho a^2}{3}$	$\dfrac{2}{5}\dfrac{2\pi a^3 \rho a^2}{5}$	$\pi a^3 \rho\left[\dfrac{a^2}{10} + \dfrac{4a^2}{15}\right]$

Hence, by the parallel axes theorem the total moment of inertia about PY is

$$I = M\left(\frac{a^2}{10} + \frac{4a^2}{15}\right) + Ma^2$$

$$= \frac{41\,Ma^2}{30}$$

(2) Derive the moment of inertia about its axis of a uniform circular annulus of mass M and internal and external radii a and $2a$. Hence, find the moment of inertia about a diameter and about a tangent to the outer rim.

The moment of inertia of the annulus about its axis is given by equations (2.28) and thus

$$I = \frac{M}{2}(4a^2 + a^2)$$

$$\therefore \qquad I = \frac{5Ma^2}{2}$$

If the moment of inertia about a diameter is I_1, then by the theorem of perpendicular axes

$$I = 2I_1 = 5\frac{Ma^2}{2}$$

$$\therefore \qquad I_1 = 5\frac{Ma^2}{4}$$

The moment of inertia about a tangent to the outer rim, I_2, is given by the theorem of parallel axes. Thus

$$I_2 = I_1 + M(2a)^2$$

$$= \frac{5Ma^2}{4} + 4Ma^2$$

$$= \frac{21Ma^2}{4}$$

2.11 ANGULAR MOMENTUM OF A RIGID BODY

With reference to *Figure 2.9*, consider a rigid body rotating with angular velocity $d\theta/dt$ about an axis through a fixed point O normal to the plane of the diagram. Let m be the mass of a small particle at a point P in the body, distance r from O. The linear velocity of the mass m will be $rd\theta/dt$ and the momentum $mrd\theta/dt$. The angular momentum about the axis of rotation will be $mr^2d\theta/dt$. Thus, the angular momentum of the whole body about the axis of rotation will be

$$\sum mr^2\dot\theta = I\dot\theta \qquad \dots (2.37)$$

where I is the moment of inertia of the body about the axis of rotation. It has previously been shown, equation (2.17), that the rate of change of angular momentum is equal to the sum of the moments of the external forces. Hence the equation of motion for a rigid body turning about a fixed axis is

$$\Gamma = I\ddot\theta = I\dot\omega \qquad \dots (2.38)$$

where Γ is the sum of the moments of the external forces about the axis, and $\omega = \dot\theta$.

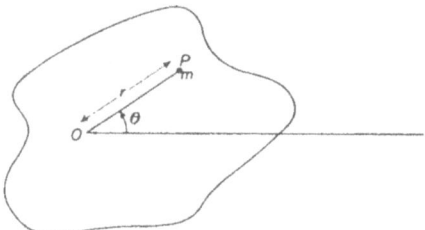

Figure 2.9. Angular momentum of a rigid body

2.12 KINETIC ENERGY OF A BODY ROTATING ABOUT AN AXIS
With reference to *Figure 2.9*, the linear velocity of the mass m at P is $r\dot\theta = r\omega$. Its kinetic energy is therefore $\frac{1}{2}mr^2\omega^2$.

The total kinetic energy of the whole body is thus

$$\sum \tfrac{1}{2}mr^2\omega^2 = \tfrac{1}{2}I\omega^2 \qquad \dots (2.39)$$

Now it has been shown that $\Gamma = I\dot\omega$, equation (2.38),

but $$\dot\omega = \frac{d\omega}{d\theta}\cdot\frac{d\theta}{dt} = \omega\frac{d\omega}{d\theta}$$

Hence $$\Gamma = I\omega\frac{d\omega}{d\theta} = \tfrac{1}{2}I\frac{d}{d\theta}(\omega^2)$$

Integrating, $$\int \Gamma d\theta = \tfrac{1}{2}I\omega^2 + A$$

40

where A is a constant of integration. If the initial angular velocity is ω_0 so that $\omega = \omega_0$ when $\theta = 0$

$$A = -\tfrac{1}{2} I \omega_0^2$$

Hence $\qquad \int \Gamma \mathrm{d}\theta = \tfrac{1}{2} I \omega^2 - \tfrac{1}{2} I \omega_0^2 \qquad \qquad \dots (2.40)$

i.e. the work done by the external forces causing the body to rotate is equal to the change in kinetic energy of the body.

2.13 DETERMINATION OF THE MOMENT OF INERTIA OF A FLYWHEEL ABOUT ITS AXIS OF ROTATION

With reference to *Figure 2.10*, the flywheel on a long axle is mounted horizontally. A known mass is attached to the axle by a cord wrapped several times around the axle. The end of the cord is passed through a hole in the axle. When the mass is allowed to fall, the potential energy lost by the mass is used in providing the

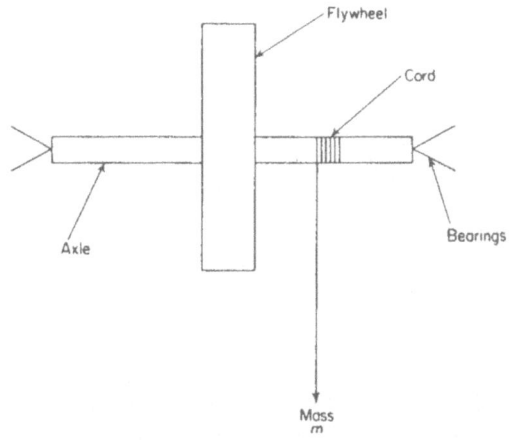

Figure 2.10. Determination of the moment of inertia of a flywheel

kinetic energy of the mass, the kinetic energy of the flywheel and in overcoming frictional forces. The length of the cord is adjusted so that as the mass reaches the floor the end of the cord leaves the axle. Let m be the mass and x the vertical distance it falls. Let I be the moment of inertia of the wheel about its axis of rotation, ω the angular velocity of the wheel when the cord leaves the axle and a the radius of the axle. Let the wheel make n_1 revolutions from rest before the mass touches the floor and let n_2 be the number of revolutions made by the wheel after the cord has left the axle. Let μ ergs of work be used in overcoming friction in each complete revolution of the wheel and let t be the time which elapses between the cord leaving the axle and the wheel coming to rest.

The energy equation for the experiment is

Potential energy lost by falling mass	=	kinetic energy gained by falling mass	+	kinetic energy gained by flywheel	+	energy used in overcoming frictional force

Therefore

$$mgx = \tfrac{1}{2}ma^2\omega^2 + \tfrac{1}{2}I\omega^2 + \mu n_1 \qquad \ldots (2.41)$$

but the maximum energy possessed by the wheel is used in overcoming frictional forces in n_2 revolutions.

Thus
$$\tfrac{1}{2}I\omega^2 = \mu n_2 \qquad \ldots (2.42)$$

Hence
$$mgx = \tfrac{1}{2}ma^2\omega^2 + \tfrac{1}{2}I\omega^2 + \frac{n_1}{n_2}(\tfrac{1}{2}I\omega^2) \qquad \ldots (2.43)$$

Now the average angular velocity of the wheel is $2\pi n_2/t$ and this is equal to half the maximum angular velocity of the wheel, assuming that frictional forces are constant. Hence

$$\omega = \frac{4\pi n_2}{t}$$

In equation (2.43), therefore, all the quantities except I may be determined and thus the value of I evaluated.

It is usual to repeat the experiment a few times with the same mass to obtain a mean value of n_2/t. Different masses are also used to provide further results.

2.13.1 Examples

(1) A heavy uniform rod AB of length $2a$ and mass M has a small mass m attached to it at B. The whole oscillates about a horizontal axis through A. Determine the time of a small oscillation.

The standard result for the moment of inertia of a rod of length l about an axis through one end perpendicular to its length is $ml^2/3$, where m is the mass of the rod. Hence, in this case the moment of inertia of the rod about A is $M4a^2/3$. The moment of inertia of the small mass about A is $m.4a^2$. Thus the total moment of inertia about A is

$$\tfrac{4}{3}Ma^2 + 4ma^2$$

Figure 2.11. Motion about a fixed axis

With reference to *Figure 2.11*, if the rod is at an angle θ to the vertical the loss in potential energy between this position and the horizontal is

$$(Mga + 2mga)\cos\theta$$

Now the gain in kinetic energy is $\tfrac{1}{2}I\omega^2 = \tfrac{1}{2}I\dot{\theta}^2$ where $d\theta/dt$ is the angular velocity at this point, i.e. gain in kinetic energy is

$$\tfrac{1}{2}(\tfrac{4}{3}Ma^2 + 4ma^2)\dot{\theta}^2$$

Hence $\quad \tfrac{1}{2}(\tfrac{4}{3}Ma^2 + 4ma^2)\dot{\theta}^2 = (Mga + 2mga)\cos\theta$

Differentiating with respect to time

$$\tfrac{1}{2}(\tfrac{4}{3}Ma^2 + 4ma^2)2\dot{\theta}\ddot{\theta} = -(Mga + 2mga)\sin\theta . \dot{\theta}$$

For small values of θ. $\sin\theta \simeq \theta$

$$\therefore \qquad (\tfrac{4}{3}Ma^2 + 4ma^2)\ddot{\theta} = -(Mga + 2mga)\theta$$

For simple harmonic motion with periodic time $T = 2\pi/\omega$, and from this equation

$$\omega = \sqrt{\left(\frac{3(Mga + 2ma)}{4(Ma^2 + 3ma^2)}\right)}$$

$$\therefore \qquad T = 4\pi\sqrt{\left(\frac{(M + 3m)\,a}{3(M + 2m)\,g}\right)}$$

43

(2) The axis of a wheel is vertical and its bearings exert on it a constant frictional couple. In one experiment it is set spinning and comes to rest in time t, during which it makes n revolutions. In a second experiment a small mass m is attached to it at a distance a from the axis, and this time, after being set spinning, comes to rest in time t', during which it makes n' revolutions. Find an expression for the moment of inertia of the unloaded wheel.

The equation of motion of the unloaded wheel subjected to a frictional couple C is

$$I\ddot{\theta} = -C$$

where I is the moment of inertia of the wheel about a vertical axis through its centre.

Integrating, $$I\dot{\theta} = -CT + b$$

where b is a constant.

At time $T = t$, $\dot{\theta} = 0$

$\therefore$ $$b = Ct$$

$\therefore$ $$I\dot{\theta} = -CT + Ct = C(t - T)$$

The initial angular velocity of the wheel $= Ct/I$ and the final angular velocity at time t is equal to zero. Thus the average angular velocity $= Ct/2I$.

Therefore the number of revolutions required to bring the wheel to rest in time t is

$$n = \frac{Ct^2}{2I}\frac{1}{2\pi} \qquad \qquad \dots (1)$$

The analysis with the wheel loaded is exactly similar except that I is replaced by $I = ma^2$.

Thus $$n' = \frac{1}{2}\frac{Ct'^2}{(I + ma^2)\,2\pi} \qquad \qquad \dots (2)$$

Eliminating C from equations (1) and (2)

$$nIt'^2 = (I + ma^2)\,n't^2$$

$\therefore$ $$I = \frac{ma^2 n' t^2}{nt'^2 - n't^2}$$

2.14 GYROSCOPIC MOTION

It has been shown previously that a body moving in a circular orbit with constant speed has a velocity at any instant along the tangent and is maintained in its circular path only because of an acceleration directed towards the centre of the circle. This acceleration constantly changes the direction of the velocity, but does not alter the speed. This is an example of uniform acceleration at constant speed. Similarly, it is possible to have angular acceleration at a constant angular speed. In such a case the plane of rotation changes direction at a given rate without any variation in the rate of rotation about the rotation axis. This type of change in the plane of rotation is called precession. It is produced by the action of a torque whose plane is perpendicular to the plane of rotation at all times. For a torque of constant magnitude constant precessional motion takes place. The magnitude of such a torque can be calculated quite simply.

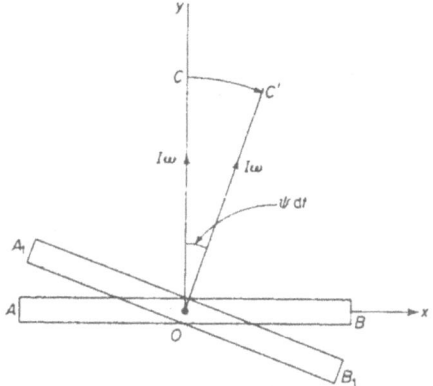

Figure 2.12. Precession of a disk

With reference to *Figure 2.12* let I be the moment of inertia of the disk about its rotation axis and $d\theta/dt = \omega$ its angular velocity. The angular momentum of the disk is thus $I\omega$. Assume the disk is rotating about the axis Oy and let the plane of the disk, AB, precess about an axis Oz perpendicular to the plane of the diagram. If the rate of precession is ψ, after a time dt the plane of the disk can be represented by A_1B_1 where the angle of precession moved through is equal to ψdt. The angular momentum, which can be represented vectorially by a line drawn normally to the plane of rotation of the disk, is represented by OC and OC' for the two positions of the disk where angle COC' is equal to ψdt. The change in angular momentum is thus represented vectorially by CC' and is thus equal to $I\omega \cdot \psi dt$. It has already been shown that the rate of change of angular

45

momentum is equal to the applied torque. Hence, if the applied torque is represented by Γ, then

$$\Gamma = \frac{I\omega \cdot \psi dt}{dt} = I\omega\psi \qquad \dots (2.44)$$

The change of angular momentum is along CC' so that it is perpendicular to the axis of rotation but parallel to the plane of rotation of the disk. Hence CC' is the axis of the applied torque Γ.

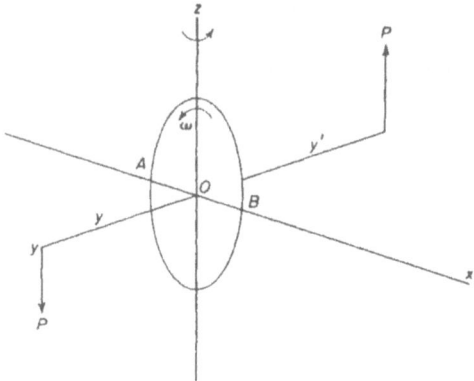

Figure 2.13. Precession of a disk

Accordingly, since the axis of rotation of the disk is Oy and the axis of the applied torque is along Ox, the disk precesses about the third mutually perpendicular axis Oz, *Figure 2.13* shows the disk being acted on by the two forces P, which constitute a couple about the Ox axis The disk precesses about the Oz axis so that A moves towards Y and B towards Y^1. The direction of precession of the disk is given by the rule which states that the direction of precession is in a direction such that a movement of the disk through $90°$ in that direction would cause the plane of the disk to be in the original plane of the couple producing the precession and such that the direction of rotation of the disk would be the same as that of the original applied torque.

2.15 GRAVITATIONAL TORQUE

Frequently, precessional motion of a rotating body is caused by the action of a gravitational torque. If the body is supported on a point,

46

which is not on the vertical through its centre of gravity, a gravitational couple acts. If the body were not rotating this gravitational couple would cause the centre of gravity of the body to be lowered. However, with a rotating body the gravitational couple provides a torque which causes the body to precess about the axis perpendicular to the axis of rotation and to the axis of the applied torque. The rate of precession, ψ, is given by equation (2.44), i.e.

$$\psi = \frac{\Gamma_1}{I\omega} \qquad \qquad \dots (2.45)$$

Where Γ_1 is the gravitational torque. This type of motion is called gyroscopic and the body is termed a gyroscope.

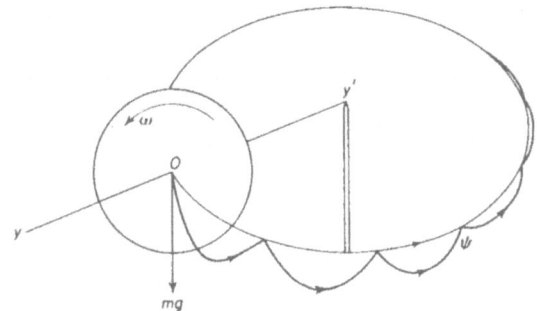

Figure 2.14. The gyrostat

Figure 2.14 illustrates an example of gyroscopic motion. The disk is rotating at an angular velocity ω about the axis YOY^1, which is supported at the point Y^1 on a vertical pivot. Let OY^1 be equal to l and the weight of the disk be equal to mg. Thus the gravitational couple acting on the disk is mgl. From equation (2.45) the precessional rate of the disk is

$$\psi = \frac{\Gamma_1}{I\omega} = \frac{mgl}{mk^2\omega} \qquad \qquad \dots (2.46)$$

where k is the radius of gyration of the disk about the axis YOY^1. It should be noted that equation (2.46) relates to the precession rate which may be maintained by the gravitational torque once it is started. However, uniform precession is not the most general motion the spinning disk can make. If a quicker precessional rate is given, then the axis YOY^1 rises, while if the precessional rate is

47

lowered YOY^1 falls. Hence the rotation axis can oscillate about the position for steady precession, and this is known as nutation.

The behaviour of the gyroscope may be considered as follows. If the part Y is held in a fixed position so that precession is not possible then there is no torque acting since the gravitational torque is balanced by the force holding Y in the fixed position. If the part Y is then released so that it is free to fall, then the gravitational torque causes the rotating disk to fall. The gyroscope is then turning, however, and for this to continue a torque is necessary. In the absence of the required torque the gyroscope begins to move in the direction opposite to that of the missing force. This gives the gyroscope a component of motion round the vertical axis. However, the actual motion becomes greater than the steady precessional rate and the axis YOY^1 rises to the level from where it started. The end of the axle Y follows a cycloid, as shown in the figure. This motion is normally rapidly damped out by friction in the gimbal bearings and only the steady precessional movement is observed.

A centrifugal torque may also act on the disk. If the centripetal reaction at the point of support is not in the same straight line as the centrifugal force, then together they constitute a couple. In the case of the disk the centripetal reaction at the point of support Y^1 is along YOY^1, while the centrifugal force is along Y^1OY and, since they act in the same straight line, they do not constitute a couple.

2.16 LANCHESTER'S RULE

If the centrifugal torque is not balanced by an equal and opposite centripetal torque the body moves outwards from the centre of the precessional motion. For balance equation (2.47) must be satisfied

$$\Gamma_1 - \Gamma_2 = \Gamma \qquad \ldots (2.47)$$

where Γ, Γ_1 and Γ_2 are the magnitudes of the gyrostatic, gravitational and centripetal torques respectively. The gravitational torque is assumed to be greater than the centripetal torque in this equation. The direction of the torque producing precession in a given direction may be determined by Lanchester's rule. This states that if a gyrostat is observed from a point in its plane with the line of sight perpendicular to the axis of precession, then a point on its circumference is observed to describe an ellipse and the motion of its path gives the direction of the precessional torque, the line of sight being its axis.

2.17 THE GYROSTATIC PENDULUM

Figure 2.15 illustrates a gyrostat revolving about a light rigid rod

AB as axis with an angular velocity ω. It is precessing about the vertical axis AO with a precessional rate ψ. Let the gyrostat have mass m and linear velocity v round the circle, centre O, at radius r. Let $AB = l$.

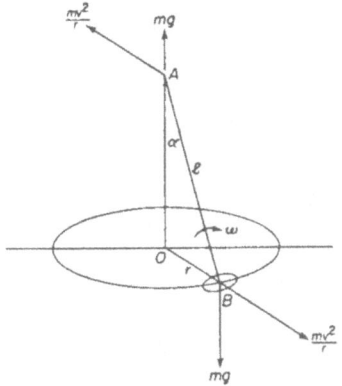

Figure 2.15. The gyrostatic pendulum

The downward force at B is mg and together with the equal upward reaction at A constitutes a couple. Thus the gravitational torque is given by

$$\Gamma_1 = mgl \sin \alpha$$

The centrifugal force on the gyrostat is mv^2/r along OB and together with an equal opposite reaction at the support constitutes a centrifugal couple equal to mv^2/r. Hence the opposing centripetal torque is given by

$$\Gamma_2 = \frac{mv^2}{r} l \cos \alpha$$

and thus acts in the same direction as Γ_1.

Since AB is perpendicular to the plane of rotation of the gyrostat, the rate of precession ψ is equal to the angular velocity of AB, v/l. Hence

$$\Gamma = I\omega\psi = mk^2\omega \frac{v}{l}$$

where k is the radius of gyration of the gyrostat about AB. Applying Lanchester's rule the direction of this gyrostatic torque is seen to be opposed to that of the gravitational and centripetal torques. Hence

$$\Gamma = -mk^2\omega\frac{v}{l}$$

when it is substituted in equation (2.47). Substituting in equation (2.47) for Γ, Γ_1 and Γ_2

$$-mk^2\omega\frac{v}{l} = mgl \sin \alpha - \frac{mv^2}{r} l \cos \alpha \qquad \dots (2.48)$$

The periodic time t is given by $vt = 2\pi r$. Substituting in equation (2.48) and simplifying

$$-k^2\omega \sin \alpha \frac{2\pi}{t} = gl \sin \alpha - l^2 \sin \alpha \cos \alpha \left(\frac{2\pi}{t}\right)^2 \qquad \dots (2.49)$$

the solution of which is

$$t = 2\pi \cdot \frac{2l^2 \cos \alpha}{k^2\omega \pm \sqrt{(k^4\omega^2 + 4gl^3 \cos \alpha)}} \qquad \dots (2.50)$$

2.18 ROLLING DISK

A further application of equation (2.47) is to the case of a disk with its plane initially vertical rolled over a horizontal surface, i.e. a coin rolled on a flat table. In such a case the coin is seen to follow a straight line path for a time until its velocity decreases to a certain value. It is then observed to lean over to one side away from the vertical and to follow a curved path on the side towards which it is leaning. The curvature of the path increases as the velocity of the coin decreases and ultimately, after describing a spiral path, the coin falls flat.

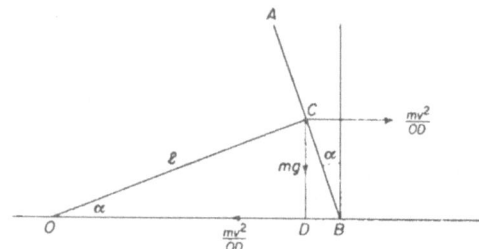

Figure 2.16. Rolling disk

With reference to *Figure 2.16*, the disk AB of radius r and mass m, is travelling with linear velocity v and its plane makes an angle α with the vertical.

The centripetal torque acting on the disk in a clockwise direction is

$$\Gamma_2 = \frac{-mv^2}{OD} \cdot CD = \frac{-mv^2 \cdot r \cos \alpha}{OC \cos \alpha} = -mv^2 \tan \alpha$$

The gravitational torque acting on the disk in a clockwise direction is

$$\Gamma_1 = -mg \cdot BD = -mgr \sin \alpha$$

and the gyrostatic torque acting on the disk in a clockwise direction is

$$\Gamma = -mk^2 \omega \psi = -mk^2 \cdot \frac{v}{r} \cdot \frac{v}{OC} = -mk^2 \cdot \frac{v^2}{r^2} \tan \alpha$$

Applying equation (2.47)

$$mk^2 \frac{v^2}{r^2} \tan \alpha = mgr \sin \alpha - mv^2 \tan \alpha \qquad \dots (2.51)$$

i.e.

$$v^2 \left(1 + \frac{k^2}{r^2}\right) = gr \cos \alpha \qquad \dots (2.52)$$

Hence, from equation (2.52) the critical velocity for straight line motion is

$$v^2 = \frac{gr}{\left(1 + \frac{k^2}{r^2}\right)} \qquad \dots (2.53)$$

2.19 GYROSCOPIC APPLICATIONS

Undoubtedly the most important application of the gyroscope is in the gyro-compass. If a gyrostat is suspended so that it has three degrees of freedom whatever the position of its support, movement of the support cannot cause a torque to act on the gyrostat and thus it stays in the same position in space. A gyrostat mounted in such a manner is called a gyro-compass and this is used for navigation in preference to the magnetic compass, particularly in submarines. The elementary gyro-compass is stable in any position and thus, after a disturbance does not return to its original direction, a disadvantage that can easily be overcome by the addition of a small weight immediately below the rotating disk of the gyro-compass, as shown in *Figure 2.17*. From this diagram it can be seen that the gyrostat is supported at the ends of its axis Y, Y' in a horizontal ring and the weight is attached to a vertical support so that it is just below the centre of the disk. The horizontal ring is mounted so that it is free to move about the axis XX' within the vertical ring which is also free to move about the axis ZZ' within the frame of the instrument. The frame is also mounted in gimbals to ensure complete freedom of movement.

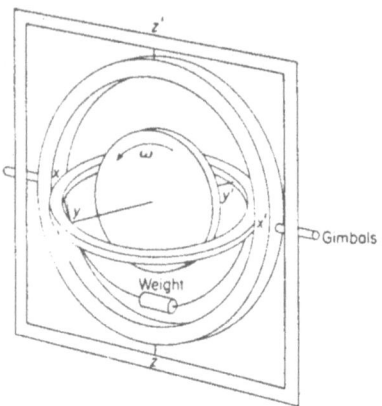

Figure 2.17. Gyro-compass

This system is only stable when the end Y of the gyrostat points north. For instance, at the equator, the axis YY' is in the north–south direction, the inner ring is horizontal and the weight is directly below the disk. Hence there is no torque due to gravity and since the axis YY' is parallel to the polar axis, the system is quite stable. If YY' were pushed into an east–west position with Y facing east as in *Figure 2.18*, then, due to the Earth's rotation, since the axis YY' is parallel to its original direction, a gravitational couple is produced due to the weight no longer being vertically below the centre of the disk. Hence precession takes place so that the end Y swings towards the north and when the axis YY' is in the north–south position the system is again stable. In a similar manner, if Y had been facing west it can be shown that the precessional motion is in the opposite direction so that Y again swings towards north.

The effect at latitudes other than the equator is more complicated, but it can be shown that the axis YY' always sets parallel to the north–south direction.

Nowadays, several other navigational aids, such as the artificial horizon, turn and bank indicators, and the automatic pilot, are also based on the gyrostat. In all these applications the gyrostat provides a fixed reference line and the movement of a line attached to the aircraft is observed relative to the reference line.

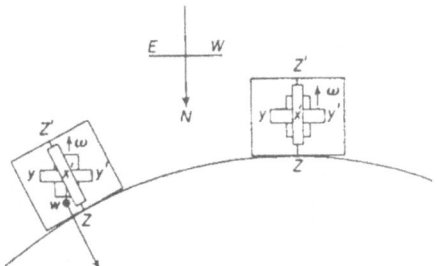

Figure 2.18. Gyro-compass at the equator

In the manufacture of rifles and artillery, it is normal to make the barrels in a fashion such that the emerging bullet or shell is spinning about an axis in its direction of motion. This spinning increases the resistance of the projectiles to small deflective forces and considerably improves their uniformity of flight.

Several other everyday applications of the gyroscopic effect exist, among them being bicycle riding and hoop rolling. In both these cases gravitational forces which tend to upset the equilibrium of the bicycle or hoop are balanced to some extent by the gyroscopic effect which changes the plane of rotation of the wheel or hoop.

3

THE ACCELERATION DUE TO GRAVITY

3.1 INTRODUCTION

THE acceleration due to gravity, denoted by the symbol g, is the acceleration of a body caused by the force acting on the body due to the earth's gravitational field. This force can be written as mg, where m is the mass of the body. Since g is the force on unit mass, then it also represents the strength of the gravitational field at the point where the measurement is made. The quantity g is independent of the mass of the body and can be accurately determined by various methods. The actual quantity measured is, in fact, the acceleration produced by the earth's attractive force minus the radial acceleration due to the earth's rotation.

Thus the direction and magnitude of the vector g are dependent on the latitude of the place where the measurement is carried out. The most accurate methods of measuring g are based upon the use of various types of pendulum, and the development of these techniques is described below.

3.2 SIMPLE PENDULUM

This consists of a light string fixed firmly at its upper end and supporting a small massive spherical body. If the body is displaced through a small distance and then released so that it oscillates in a vertical plane it is said to describe simple harmonic motion. With reference to *Figure 3.1*, if l is the length of the string, θ the displacement of the body at any instant of time and m the mass of the body, then the restoring force acting on the body is $mg \sin \theta$. Thus the acceleration of the body towards the centre is $g \sin \theta$, which may be approximated to $g\theta$ for small values of θ. The velocity of the body in this position is $l\dot{\theta}$ and its acceleration towards the centre is therefore $-l\ddot{\theta}$, hence

$$g\theta = -l\ddot{\theta}$$

which may be written as

$$\ddot{\theta} + \frac{g}{l}\theta = 0 \qquad \qquad \dots (3.1)$$

54

This equation represents simple harmonic motion of period given by

$$T_0 = 2\pi\sqrt{\frac{l}{g}} \qquad \dots (3.2)$$

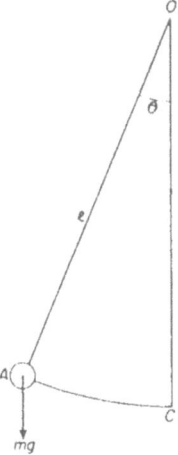

Figure 3.1. The simple pendulum

3.2.1. Corrections applicable to the use of the simple pendulum

In the derivation of equation (3.2) it was assumed that the string was of negligible mass, the oscillations were of infinitely small amplitude, the body was very small and that the body was oscillating *in vacuo*. These conditions are not fulfilled practically and it is necessary to make allowances for the finite amplitude of the swing, the finite size of the body and for the viscous drag of the medium in which the body is oscillating.

Consider first of all the effect of the viscous retarding force acting on the body. It can be shown that the retarding force is proportional to the linear velocity of the body $l\dot{\theta}$ and to the viscosity of the medium.

The equation of motion of the pendulum about the axis through O becomes

$$ml^2\ddot{\theta} + kl\dot{\theta} + mgl\theta = 0 \qquad \dots (3.3)$$

If $k/ml = 2b$ and $g/l = c^2$ this equation becomes

$$\ddot{\theta} + 2b\dot{\theta} + c^2\theta = 0 \qquad \dots (3.4)$$

and the general solution of this may be written in the form

$$\theta = Ae^{-bt} \cos\left[(c^2 - b^2)^{\frac{1}{2}} t + \phi\right] \qquad \ldots (3.5)$$

This represents damped oscillations whose period is given by the equation

$$T = \frac{2\pi}{(c^2 - b^2)^{\frac{1}{2}}} = \frac{2\pi}{\left(\dfrac{g}{l} - b^2\right)^{\frac{1}{2}}} \qquad \ldots (3.6)$$

i.e.

$$T = T_0 \left[\frac{g}{l\left(\dfrac{g}{l} - b^2\right)}\right]^{\frac{1}{2}} \qquad \ldots (3.7)$$

For a pendulum vibrating in air, then the term b is small and equation (3.7) may be written as

$$T = T_0 \left(1 + \frac{b^2 l}{2g}\right)$$

i.e.

$$T_0 = T \left(1 - \frac{b^2 l}{2g}\right) \qquad \ldots (3.8)$$

The variation of the angular displacement θ of the body with time t is illustrated in *Figure 3.2*. Although for a body oscillating in

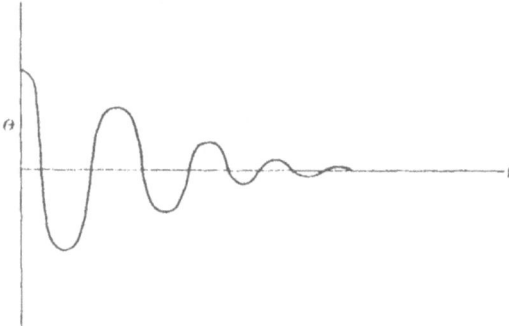

Figure 3.2. Damped oscillations

air the viscous drag of air on the body is very small, for accurate measurements this gradual decay of the oscillations must be taken into account.

When the angular displacement θ is large, so that it is not permissible to approximate $\sin \theta$ to θ, equation (3.1) becomes modified to

$$\ddot{\theta} + \frac{g}{l}\sin \theta = 0 \qquad \dots (3.9)$$

A solution of this equation is provided in several standard text books, e.g. Sharman, 'Vibration and Waves' Section 3.1. and the resulting period of oscillation is found to be given by

$$T = 2\pi k(\theta)\sqrt{\frac{l}{g}} \qquad \dots (3.10)$$

In this equation $k(\theta)$ is a function of θ and is a complete elliptic integral of the first kind. Expansion of this function gives the equation

$$T = 2\pi\sqrt{\frac{l}{g}}\left[1 + \left(\frac{1}{2}\right)^2 \sin^2 \frac{\theta}{2} + \left(\frac{1.3}{2.4}\right)^2 \sin^4 \frac{\theta}{2} + \dots\right]$$

$$\dots (3.11)$$

If θ is small, $\sin \theta/2$ may be replaced by $\theta/2$ and subsequent terms may be neglected. Hence equation (3.5) becomes

$$T = 2\pi\sqrt{\frac{l}{g}}\left(1 + \frac{\theta^2}{16}\right) = T_0\left(1 + \frac{\theta^2}{16}\right) \qquad \dots (3.12)$$

An approximate correction for the finite size of the oscillating body can be made as follows. Assume the body is spherical, has mass m, and is of radius a. Assume also that the body moves so that the radius AB of the body, *Figure 3.3*, always lies along the line AO joining the centre of the body to the point of support O. Hence the body oscillates about an axis through O, perpendicular to the plane of the diagram. Now it has previously been shown that the moment of inertia of a sphere about a diameter is $2ma^2/5$. Hence in this case, applying the theorem of parallel axes, the moment of inertia of the body about an axis parallel to a diameter and at a distance l from it is

$$\frac{2ma^2}{5} + ml^2$$

Thus, the equation of rotational motion for small oscillations of the body, neglecting viscous forces, is

57

$$(\tfrac{2}{5}ma^2 + ml^2)\,\ddot\theta + mgl\theta = 0 \qquad \ldots (3.13)$$

and the period of oscillation is given by

$$T_1 = 2\pi \sqrt{\frac{2a^2/5 + l^2}{lg}} \qquad \ldots (3.14)$$

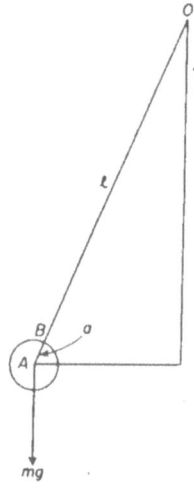

Figure 3.3. Pendulum with large bob

Naturally, the body does not oscillate in the manner assumed, but oscillates about the point B after the string has reached its maximum amplitude on each side. In addition, the above three corrections do not take into account the finite mass of the supporting string. In view of these and other factors which are not corrected for, the value of g obtained by use of the simple pendulum is not very accurate. Accurate methods for the measurement of g are mainly based on the use of a compound pendulum which is discussed later.

3.2.2. Examples
(1) A simple pendulum consists of a bob, of mass m, suspended from a fixed point O by an inextensible string of length l. The pendulum oscillates in a vertical plane and when the string is inclined at an angle θ to the vertical, its angular velocity is ω and its tension is T. Prove that

$$\omega^2 = \frac{2g}{l}\cos\theta + A$$

and $$T = 3mg\cos\theta + B$$

where A and B are constants.

Find the angular acceleration of the string as a function of θ. If θ is small, show that the tension is approximately constant throughout the motion and that the motion of the bob is approximately simple harmonic.

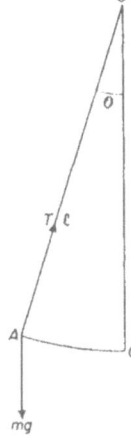

Figure 3.4. Simple pendulum problem

With reference to *Figure 3.4*, the bob has a component of acceleration directed towards O along the string. Hence

$$ml\dot{\theta}^2 = T - mg\cos\theta \qquad \ldots (1)$$

Let the kinetic energy of the bob as it passes through C equal C'. Applying the principle of conservation of mechanical energy

$$C' - \tfrac{1}{2}ml^2\dot{\theta}^2 = mgl(1 - \cos\theta) \qquad \ldots (2)$$

N.B. linear velocity of bob $= l\dot{\theta}$
Hence

$$\dot{\theta}^2 = \frac{2C'}{ml^2} - \frac{2g}{l} + \frac{2g}{l}\cos\theta$$

$$\therefore \qquad \dot{\theta}^2 = \omega^2 = A + \frac{2g}{l}\cos\theta \qquad \ldots (3)$$

where $A = \dfrac{2C'}{ml^2} - \dfrac{2g}{l}$ and is a constant for the system.

Substituting for $\dot\theta^2$ in equation (1),

$$ml\left(A + \frac{2g}{l}\cos\theta\right) = T - mg\cos\theta$$

$\therefore$
$$T = mlA + 3mg\cos\theta$$
$$= B + 3mg\cos\theta \qquad \dots (4)$$

Differentiating equation (3) gives

$$2\dot\theta\ddot\theta = -\frac{2g}{l}\sin\theta\,.\,\dot\theta$$

$\therefore$
$$\ddot\theta = -\frac{g}{l}\sin\theta$$

Therefore, for small values of θ

$$\ddot\theta = -\frac{g}{l}\theta$$

which is the equation of simple harmonic motion of period $2\pi\sqrt{(l/g)}$.
From equation (4), for small values of θ

$$T = B + 3mg$$

and the tension is approximately constant throughout the motion.

(2) A pendulum which beats seconds at a place where $g = 32\cdot18$ ft/s^{-2} is taken to a place where it loses 210 s/day. Find the value of g at the latter place.

$$T = 2\pi\sqrt{\frac{l}{g}}$$

From the general theory of errors

$$\frac{dT}{T} = \frac{1}{2}\frac{dl}{l} - \frac{1}{2}\frac{dg}{g}$$

$\therefore$
$$\frac{2dT}{T} = \frac{dl}{l} - \frac{dg}{g}$$

In this problem, $T = 2$ and $dl = 0$.

$\therefore$ $\qquad dT = 210\,.\,2/86{,}400$ s/period

and $\quad$ $dg = -210.2.32{\cdot}18/86{,}400 = -0{\cdot}156$

Thus value of g at the new place

$$= 32{\cdot}18 + dg$$
$$= 32{\cdot}18 - 0{\cdot}156$$
$$= 32{\cdot}024 \text{ ft s}^{-2}$$

3.3 COMPOUND PENDULUM

The compound pendulum consists of a rigid body which oscillates about a fixed horizontal axis. *Figure 3.5* shows such a rigid body which is free to oscillate about a horizontal axis through O perpendicular to the plane of the diagram. Let A represent the centre of gravity of the body when it is displaced through an angle θ at any instant of time. Let the distance from the centre of gravity to the point of suspension, AO, be l and let the mass of the body be

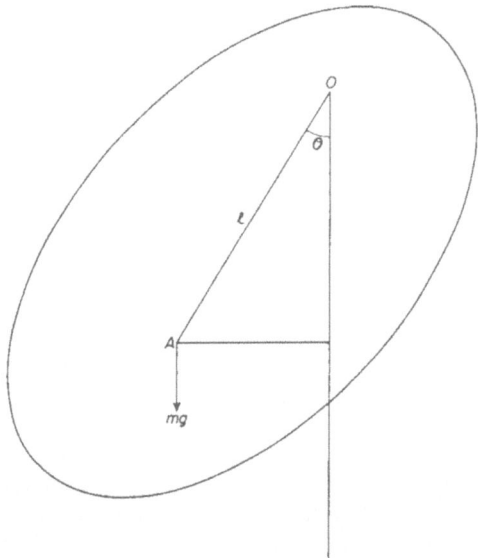

Figure 3.5. Compound pendulum

m. If the moment of inertia of the body about an axis through A is mk^2, then the moment of inertia about a parallel axis through O is, by the theorem of parallel axes, $mk^2 + ml^2$. If the displacement θ is assumed to be small, then the equation of rotational motion about the axis through O is

61

$$(mk^2 + ml^2)\ddot{\theta} + mgl\theta = 0 \qquad \dots (3.15)$$

The period of rotation is given by

$$T_0 = 2\pi \sqrt{\left(\frac{k^2 + l^2}{lg}\right)} \qquad \dots (3.16)$$

If l is varied, and the period of oscillation determined in each case, then the graph of T_0 against l has the form shown in *Figure 3.6*.

In practice, the compound pendulum normally consists of a uniform metal bar with a regular series of holes drilled in it. The bar is supported on a knife edge which is placed through each of the holes in turn and the period T_0 measured in each case. The distance l from the centre of gravity of the bar to the point of suspension is measured in each case and hence a graph of T_0 against l may be plotted. The centre of gravity of the bar may be conveniently determined by balancing it on a knife edge.

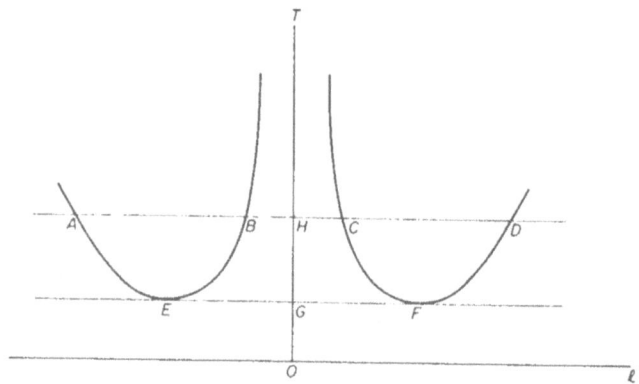

Figure 3.6. Compound pendulum—variation of T with l

As may be seen from *Figure 3.6*, it is found that there are two values of l when the support is on the same side of the centre of gravity of the bar for which the period is the same. With reference to the diagram let $HC = l_1$ and $HD = l_2$. Hence if the period of oscillation for these lengths is T, then

$$T = 2\pi \sqrt{\left(\frac{k^2 + l_1^2}{l_1 g}\right)} \qquad \dots (3.17)$$

and

$$T = 2\pi \sqrt{\left(\frac{k^2 + l_2^2}{l_2 g}\right)} \qquad \dots (3.18)$$

62

i.e.
$$\frac{gl_1T^2}{4\pi^2} = k^2 + l_1^2$$

and
$$\frac{gl_2T^2}{4\pi^2} = k^2 + l_2^2$$

Subtracting, there results

$$T = 2\pi\sqrt{\frac{l_1 + l_2}{g}} \qquad \qquad \dots (3.19)$$

It can also be seen that on the other side of the centre of gravity there are a further two points, A and B, where the period is the same. $HB = HC = l_1$ and $HA = HD = l_2$. Hence $BD = AC = l_1 + l_2$, i.e. B and D or A and C are two points whose distance apart is equal to the length of the simple pendulum whose period of oscillation is the same as that of the body when it is suspended from either B, D, A or C.

A minimum period of oscillation of the compound pendulum may be obtained when the body is suspended from two points represented by E and F, situated at equal distances from the centre of gravity such that $GE = GF = l_0$.

Differentiation of equation (3.16) with respect to l shows that for a minimum value of T, i.e.

$$\frac{dT}{dl} = 0$$

$$l = k$$

Hence, if the period of vibration at E and F is $T_{min.}$, the distance $EF = 2l_0 = 2k$, and thus

$$T_{min.} = 2\pi\sqrt{\frac{2k}{g}} \qquad \qquad \dots (3.20)$$

so that the body acts as a simple pendulum of length $2k$.

In the determination of g by the compound pendulum, after the graph of *Figure 3.6* has been plotted, lines such as $A\,B\,H\,C\,D$ are drawn and an average value is taken of the distances AC and BD, thus enabling an approximate average value of g to be obtained.

3.4 KATER'S REVERSIBLE PENDULUM
Several accurate determinations of g have been made by various types of compound pendulum and one of the most famous is the pendulum designed and used by Kater in 1817. With reference to *Figure 3.7*, it consists of a bar weighted at one end so that the

63

centre of gravity is nearer to one end than the other. The bar is provided with two knife edges K_1 and K_2, one near to either end of the bar. Two adjustable masses A and B can be moved along the bar between the knife edges. To facilitate adjustments the bar has a scale engraved upon it. The periods of oscillation about the two knife edges are measured and the two weights A and B are adjusted until these two periodic times are equal. Usually one of the weights is much smaller than the other in order to afford a fine adjustment.

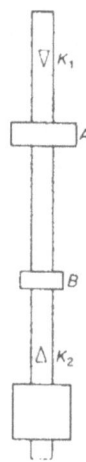

Figure 3.7. Kater's pendulum

In the original work of Kater the pendulum was adjusted so that the two periodic times were equal to within a fraction of one vibration in 24 h. However, in present-day experiments, it is usual to make the times only approximately equal. In this case, if the periodic times about the knife edges K_1 and K_2 are T_1 and T_2 and if K_1 and K_2 are distance l_1 and l_2 respectively from the centre of gravity of the pendulum, then

$$T_1 = 2\pi\sqrt{\left(\frac{k^2 + l_1^2}{gl_1}\right)} \text{ and } T_2 = 2\pi\sqrt{\left(\frac{k^2 + l_2^2}{gl_2}\right)}$$

Hence
$$\frac{g}{4\pi^2}(T_1^2 l_1 - T_2^2 l_2) = l_1^2 - l_2^2$$

Rearranging

64

$$\frac{4\pi^2}{g} = \frac{T_1^2 l_1 - T_2^2 l_2}{l_1^2 - l_2^2} = \frac{1}{2}\left(\frac{T_1^2 + T_2^2}{l_1 + l_2} + \frac{T_1^2 - T_2^2}{l_1 - l_2}\right) \quad \ldots (3.21)$$

If T_1 is approximately equal to T_2 then the term $(T_1^2 - T_2^2)/(l_1 - l_2)$ is very small compared with the term $(T_1^2 + T_2^2)/(l_1 + l_2)$ and does not need to be determined exactly. The distance between the knife edges K_1 and K_2 gives $l_1 + l_2$, while the centre of gravity of the pendulum may be determined by balancing it on a knife edge.

Various techniques have been developed in order to determine the periodic times T_1 and T_2 as accurately as possible and details of such methods are provided in textbooks of experimental physics.

3.5 CORRECTIONS APPLICABLE TO THE COMPOUND PENDULUM

For a really accurate determination of g, several corrections must be applied when the compound pendulum is used:

(1) In the first place, just as with the simple pendulum, equation (3.16) was derived on the assumption that the amplitude of oscillation is extremely small. If θ is not very small, then equation (3.15) must be modified by using $\sin\theta$ for θ. This equation can then be solved in a similar manner to the solution of equation (3.9) which was discussed earlier. The result gives the following equation for the periodic time measured:

$$T = 2\pi\sqrt{\frac{k^2 + l^2}{lg}} \cdot \left(1 + (\tfrac{1}{2})^2 \sin^2\frac{\theta}{2} + \ldots\right) \quad \ldots (3.22)$$

and if θ is small, this equation becomes

$$T = T_0\left(1 + \frac{\theta^2}{16}\right) \quad \ldots (3.23)$$

where
$$T_0 = 2\pi\sqrt{\frac{k^2 + l^2}{lg}}$$

Thus in order to obtain T_0, the measured period T must be multiplied by the factor

$$1 - \frac{\theta^2}{16}$$

since
$$T_0 = T\left(1 - \frac{\theta^2}{16}\right) \quad \ldots (3.24)$$

(2) The fact that the pendulum is oscillating in air and not *in vacuo* gives rise to several errors. Corrections must be made for the

65

viscous resistance of the air, for the upthrust of the air displaced by the pendulum, and for the fact that the pendulum drags air along with it during its motion and gives rise to an increased effective mass.

The effect of the viscous retarding force acting on the pendulum has previously been discussed in Subsection 3.2.1 in connection with the oscillations of the simple pendulum. The result is to cause the measured period T to slightly exceed T_0 where

$$T_0 = 2\pi \sqrt{\frac{k^2 + l^2}{lg}}$$

The upthrust of the air displaced by the pendulum causes the effective weight of the pendulum mg to be slightly reduced and this also causes the measured period T to slightly exceed T_0. The simple equation (3.15) becomes modified to

$$(mk^2 + ml^2)\ddot{\theta} + (m - m_1)gl\theta = 0 \qquad \dots (3.25)$$

where m_1 is the mass of air displaced by the pendulum. This mass may be calculated from the volume of the pendulum and the density of the air.

The measured period of oscillation T is thus given by the equation

$$T = 2\pi \sqrt{\frac{k^2 + l^2}{lg\left(1 - \frac{m_1}{m}\right)}} \qquad \dots (3.26)$$

The air dragged along by the pendulum during its motion gives rise to an increased effective mass which thus increases the effective moment of inertia. This additional term also causes the measured periodic time to be slightly greater than T_0. Assuming the mass of air dragged along by the pendulum to equal m_2, the moment of inertia of the system about the axis of rotation will be given by $mk^2 + ml^2 + m_2d^2$, where m_2d^2 represents the increase in the moment of inertia due to the air. The equation of motion then becomes

$$(mk^2 + ml^2 + m_2d^2)\ddot{\theta} + (m - m_1)gl\theta = 0 \qquad \dots (3.27)$$

and the periodic time is given by

$$T = 2\pi \sqrt{\frac{k^2 + l^2 + \frac{m_2}{m}d^2}{lg\left(1 - \frac{m_1}{m}\right)}} \qquad \dots (3.28)$$

The effect of the upthrust of the air was first taken into account by Newton, while the effect of the air dragged along was first investigated by Du Buat. For this reason the latter correction is often referred to as Du Buat's correction. At a later date Bessel investigated the effects more thoroughly and showed that the elementary treatment presented above is not completely accurate. In modern work involving the compound pendulum the effects produced by the pendulum oscillating in air are reduced to negligible proportions by carrying out the experiments under a reduced pressure. It has been established that at low pressures the effects are a linear function of pressure and hence if measurements are carried out at a few different pressures the results may be extrapolated to give the periodic time at zero pressure.

(3) A further correction could be necessary, due to possible curvature of the knife edges of the pendulum. In modern work, however, this effect is normally avoided by having plane bearings accurately ground flat on the pendulum and a fixed knife edge, ground to a sharp edge, on the support. The plane bearings are always replaced in the same position on the knife edge.

(4) A final correction is necessary to take into account the yielding of the support, since unless it is very rigidly fixed, the support will oscillate co-periodically with the pendulum.

With the reference to *Figure 3.8(a)*, let O be the point of suspension of the pendulum and A its centre of gravity. The components of the acceleration of A along OA and normal to it are $l_1\dot\theta^2$ and $l_1\ddot\theta$ respectively. Thus for small values of θ, the horizontal acceleration of A approximately equals $l_1\ddot\theta$ while its vertical acceleration is $l_1\dot\theta^2$. Hence, the force acting in the horizontal direction is $m_1 l_1\ddot\theta$; but from equation (3.15)

$$\ddot\theta = -\frac{g l_1 \theta}{k^2 + l_1^2}$$

and thus, the force acting on the support in the horizontal direction is given by

$$X = \frac{mg l_1^2 \theta}{k^2 + l_1^2}$$

$$= \frac{mg l_1 \theta}{l_1 + l_2^\eta} . \qquad \dots (3.29)$$

since $\qquad k^2 = l_1 l_2^\eta$

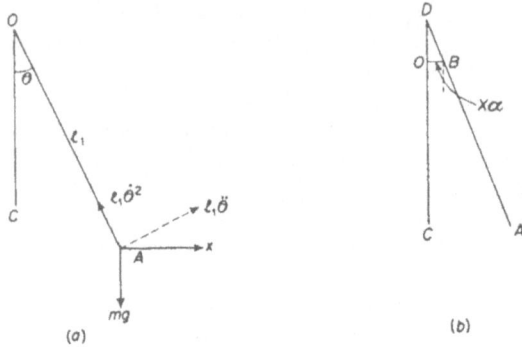

Figure 3.8. Yielding of pendulum support

If the support yields in the horizontal direction by an amount α per unit force, then the yield represented by the distance OB in *Figure 3.8(b)* is $X\alpha$, therefore

$$OB = \frac{mgl_1\theta\alpha}{l_1 + l_2} \qquad \cdots (3.30)$$

The new position of the pendulum is represented by AB, and D, the instantaneous centre of its motion is given by the intersection of AB with the vertical through O. For small values of θ, BD is approximately equal to OD and since $OB = OD . \theta$, then the effective distance of A from the axis of rotation is increased by BD (δ_1) where

$$BD = \frac{mgl_1\alpha}{l_1 + l_2} = \delta_1$$

If the period for the pendulum is T_1, then

$$\frac{gT_1^2}{4\pi^2} = \frac{(l_1 + \delta_1)^2 + k^2}{(l_1 + \delta_1)} = l_1 + \delta_1 + \frac{k^2}{l_1 + \delta_1} \qquad \cdots (3.31)$$

Similarly for the other knife edge

$$\frac{gT_2^2}{4\pi^2} = l_2 + \delta_2 + \frac{k^2}{l_2 + \delta_2} \qquad \cdots (3.32)$$

68

Hence, combining equations (3.31) and (3.32)

$$\frac{g}{4\pi^2}\left(\frac{T_1^2 l_1 - T_2^2 l_2}{l_1 - l_2}\right) = \frac{1}{l_1 - l_2}\left(l_1^2 + \delta_1 l_1 + \frac{k^2 l_1}{l_1 + \delta_1}\right)$$

$$- l_2^2 - l_2 \delta_2 - \frac{k^2 l_2}{l_2 + \delta_2}\right) \simeq (l_1 + l_2) + \frac{\delta_1 l_1 - \delta_2 l_2}{l_1 - l_2}$$

Substituting for δ_1 and δ_2

$$\frac{g}{4\pi^2}\left[\frac{T_1^2 l_1 - T_2^2 l_2}{l_1 - l_2}\right] = l_1 + l_2 + mg\alpha \quad \dots (3.33)$$

Thus the yielding of the support increases the effective length of the simple equivalent pendulum and the measured period must be reduced by the appropriate amount to obtain T_0. The correction factor of $mg\alpha$ is the movement of the support which is produced by a horizontal force equal to the weight of the pendulum. This can easily be determined by hanging the pendulum over a pulley by a string attached horizontally to the support. The amount of yield can be directly measured by a travelling microscope. The vertical force acting on the support need not be considered here, since the acceleration in the vertical direction of the pendulum is of the second order in θ and the force will have a very small effect compared with the horizontal force.

3.5.1. Example

A uniform rod AB of mass M and length a is rigidly attached at B to a point on the rim of a uniform circular disk of radius a and mass $2M$, the rod being in the direction of the outward drawn radius through B. Find the moment of inertia of the system about the axis through A perpendicular to the plane of the disk; the system hangs freely from A. Write down the equation of motion for small oscillations in the vertical plane of the disk and show that the time period is the same as that of a simple pendulum of length $56a/27$.

With reference to *Figure 3.9(a)*, the moment of inertia of the disk about a perpendicular axis through $0 = \frac{1}{2}.2Ma^2$. Its moment of inertia about a parallel axis through A by the theorem of parallel axes $= Ma^2 + 2M(2a)^2 = 9Ma^2$. The moment of inertia of rod AB about a perpendicular axis through A

$$= \frac{4M}{3}\left(\frac{a}{2}\right)^2 = \frac{Ma^2}{3}$$

69

D

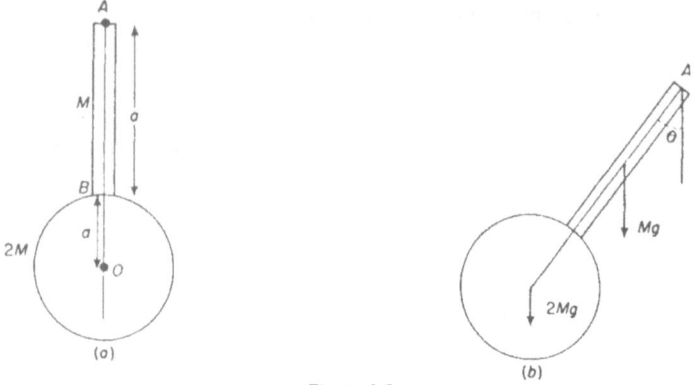

Figure 3.9

Hence the total moment of inertia of the whole body about a perpendicular axis through $A = 28Ma^2/3$.

The equation of motion of the body oscillating about A, at any time, t, is

$$I\ddot{\theta} = -Mg\frac{a}{2}\sin\theta - 2Mg\,2a\sin\theta$$

$$= -\frac{9}{2}Mga\sin\theta$$

where I is its moment of inertia about a perpendicular axis through A

$$\therefore \qquad \ddot{\theta} = -\frac{9}{2}Mga\sin\theta \Big/ \frac{28Ma^2}{3}$$

$$= -\frac{g}{a}\cdot\frac{27}{56}\cdot\sin\theta$$

For small oscillations $\sin\theta \sim \theta$ and

$$\ddot{\theta} = -\frac{g}{a}\cdot\frac{27}{56}\cdot\theta$$

Thus the motion is simple harmonic of period $2\pi\sqrt{(56a/27g)}$ and the length of the simple equivalent pendulum is $56a/27g$.

3.6 VARIATION OF THE ACCELERATION DUE TO GRAVITY

It has already been pointed out that the quantity g measured is, in fact, the acceleration produced by the earth's attractive force minus the radial acceleration due to the earth's rotation, i.e. the attraction of the earth produces two effects:

(1) It supplies the centripetal force mv^2/R necessary to maintain the body at a constant distance from the centre of the earth.
(2) It produces the acceleration due to gravity which is measured.
   The first effect varies with the latitude of the place where the measurement is carried out and can be allowed for as follows.

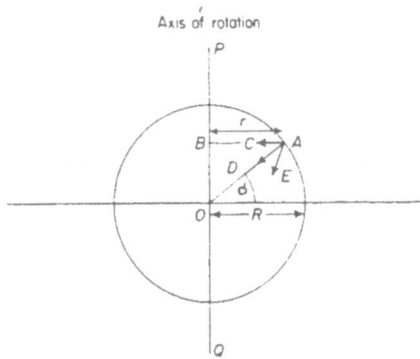

Figure 3.10. Variation of gravity with latitude

   With reference to *Figure 3.10*, suppose a body of mass m is situated at A on the earth's surface, where A is at latitude ϕ. The gravitational attraction on the body is of magnitude mg and acts along AO. It is represented by the vector AD. Here g is used to represent the value of the acceleration due to gravity which would be measured at A if the earth were stationary. Now due to the rotation of the earth the body moves in a circle of radius r and thus a centripetal force must be exerted on the body of magnitude mv^2/r where v is the linear velocity of the body. This force acts along AB and is represented by the vector AC. Thus the apparent gravitational force at A is given by the difference in the vectors AD and AC. Vectorially this is done by drawing a vector DE equal to AC, but in the opposite direction, and then adding it to AD. The resultant vector AE then represents the apparent gravitational force on the body of A and it shows the direction in which a plumb line suspended at A would come to rest. It is important to realize that the direction of the plumb line is perpendicular to the actual surface of the earth since the earth's surface is everywhere perpendicular to the apparent gravitational force. Hence the earth has a shape whereby it is flattened at the poles and bulging at the equator.

71

The magnitude of the apparent gravitational force can be simply calculated vectorially as follows:

$$AE^2 = AD^2 + DE^2 - 2AD . DE \cos \phi \qquad \ldots (3.34)$$

If g_ϕ is the measured value of the acceleration due to gravity at A, then

$$(mg_\phi)^2 = (mg)^2 + \left(\frac{mv^2}{r}\right)^2 - 2mg . \frac{mv^2}{r} \cos \phi$$

Since $v^2/r = r\theta^2$ and $r = R \cos \phi$, where θ is the angular velocity of rotation of the earth, then

$$(mg_\phi)^2 = (mg)^2 + (m\theta^2 R \cos \phi)^2 - 2mg . m\theta^2 R \cos^2 \phi$$

$$\therefore \quad g_\phi^2 = g^2 + \theta^4 R^2 \cos^2 \phi - 2g\theta^2 R \cos^2 \phi \qquad \ldots (3.35)$$

In this expression g is the theoretical value of the acceleration due to gravity which would be measured if the earth were not rotating. It cannot be measured directly, of course, and it is necessary to eliminate it from equation (3.35). At the equator $\phi = 0$, and thus

$$g_\phi^2 = g^2 + \theta^4 R^2 - 2g\theta^2 R$$

i.e.
$$g_0^2 = (g - \theta^2 R)^2 \qquad \ldots (3.36)$$

where g_0 is the value measured at the equator. Hence

$$g = g_0 + \theta^2 R \qquad \ldots (3.37)$$

Substituting equation (3.37) into equation (3.35)

$$g_\phi^2 = (g_0 + \theta^2 R)^2 + \theta^4 R^2 \cos^2 \phi - 2(g_0 + \theta^2 R) \theta^2 R \cos^2 \phi$$

Simplifying and rearranging

$$\therefore \qquad g_\phi^2 = g_0^2 + \theta^2 R(2g_0 - \theta^2 R) \sin^2 \phi \qquad \ldots (3.38)$$

Equation (3.38) provides a relationship between the value of the measured acceleration due to gravity at any latitude ϕ to the measured value obtained at the equator. This equation was derived taking into account the rotation of the earth; however, the derivation did not take into account the ellipticity of the earth. If this factor is considered, then the radius R varies with latitude and the resulting equation relating g_ϕ and g_0 is

$$g_\phi = g_0(1 + k \sin^2 \phi) \qquad \text{....(3.39)}$$

where k is a constant.

3.7 GRAVITY SURVEYS

A gravity survey is carried out in order to compare the values of the acceleration due to gravity g at various points in a country with the value of g at some chosen standard position, say g_0. Originally this type of survey was carried out by using the reversible pendulums and making an absolute determination of the value of g at each place. Since the use of the reversible pendulum is extremely laborious, present-day gravity surveys are carried out with the aid of instruments which enable relative measurements of g to be made with respect to its value at some chosen point. For instance, if a pendulum is used for such relative measurements, then if the periodic time at the chosen standard position is T_0

$$T_0 = 2\pi \sqrt{\frac{l}{g_0}} \qquad \text{....(3.40)}$$

where l is the length of the simple equivalent pendulum. If, at any other point the periodic time using the same pendulum is T_1, then

$$T_1 = 2\pi \sqrt{\frac{l}{g}} \qquad \text{....(3.41)}$$

Hence, combining equations (3.40) and (3.41)

$$g = g_0 \frac{T_0^2}{T_1^2} \qquad \text{....(3.42)}$$

which gives the value of g in terms of g_0.

Various instruments for the measurement of relative values of g have been developed and some of the more important are discussed below.

3.8 THE INVARIABLE PENDULUM

This consists of a rigid pendulum made of the nickel-steel alloy 'Invar' whose coefficient of linear expansion with temperature is very small. The pendulum has an agate knife edge which rests on an agate plane mounted on a massive support. The entire apparatus is placed inside a vessel which can be evacuated, thereby eliminating several corrections. If the pendulum oscillates in various places then the values of g obtained at the various points are related to the value of g at the standard point by equation (3.42). The value of

g at the standard point can be determined absolutely by the reversible pendulum. This equipment affords an excellent method for measuring the variation of g since the only factor which affects the periodic time, besides the change in g, is the variation in temperature, which affects the length of the pendulum. This effect is made small by the use of Invar and in addition, it is a simple matter to measure the variation of periodic time with temperature and thus to correct all the measurements to some standard temperature. A correction needs to be made for the finite amplitude of swing of the pendulum, but this is usually very small because the amplitude is of the order of one or two degrees.

The accuracy of the measurements is governed by the accuracy with which the periodic time is measured, and this can be done to a high order of accuracy by sending wireless signals to the place where the measurement is being carried out from a base station which is equipped with a standard clock.

3.9 GRAVITY METERS

Gravity meters are instruments which enable relative measurements of g to be made, but not to the same order of accuracy as the invariable pendulum. They are often used for prospecting work where

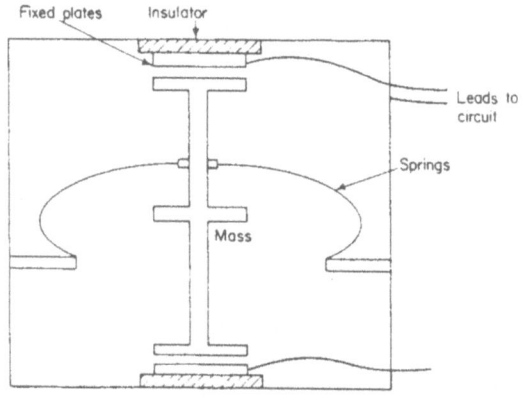

Figure 3.11. Boliden gravity meter

abnormal conditions may exist due to very high or very low density constituents in the earth's surface. Several different types of gravity meter are in use and one or two examples are described below.

74

The *Boliden gravity meter*, which consists of a mass bounded by two flat surfaces, is illustrated diagrammatically in *Figure 3.11*. Its body is supported by the springs shown. The flat plates are parallel to and close to the fixed plates which are insulated from the framework of the equipment. The two pairs of plates thus form parallel plate condensers. The upper condenser forms part of an oscillatory circuit, the frequency of which is checked against a standard oscillator. The gap between the condenser plates varies with the value of the gravitational field and thus the capacity of the upper condenser varies with the value of g. If c is the value of the capacitance at some standard position for a plate separation x, then if the plate separation changes by a distance δx the change in capacitance δc is given by

$$\delta c = \frac{c\delta x}{x} \qquad \qquad \dots (3.43)$$

If the change in g is δg, then

$$\delta g = k_1\delta x = k_2\delta c = k_3\delta N \qquad \dots (3.44)$$

where δN is the change in frequency produced in the oscillator and k_1, k_2 and k_3 are constants. Once the meter is calibrated, the frequency change affords a measure of the variations of δg directly. One method of calibration consists of applying a known potential to the lower condenser plates and measuring the frequency change produced, the attractive force between the plates being calculated. A better way of using the instrument is by applying a known potential difference to either of the sets of condenser plates in order to counterbalance the frequency change produced by the variation in g. In this method, when the frequency has been adjusted to its original value at the standard position, the attractive force can again be calculated. This type of gravity meter is sensitive to about 1 milligal, where a milligal is defined by the relationship

$$1 \text{ cm s}^{-2} = 1000 \text{ milligals}$$

The milligal is the usual unit used when small variations in g are being measured.

The *Hartley gravity meter* is another type of instrument which affords a reliable method for measuring variation of g and has the same order of sensitivity as the Boliden meter. The instrument is illustrated in *Figure 3.12* and consists of a mass suspended by a spring. A light scale beam is attached between the spring and the load and it is hinged at one end to a fixed support, while the other end carries two mirrors. The Hartley gravimeter is based on the principle that if an upward force is applied to a suspended mass, so that the force is not quite equal to the gravitational attraction, then it is a relatively simple matter to measure changes in the small additional force which is required to hold the system in equilibrium.

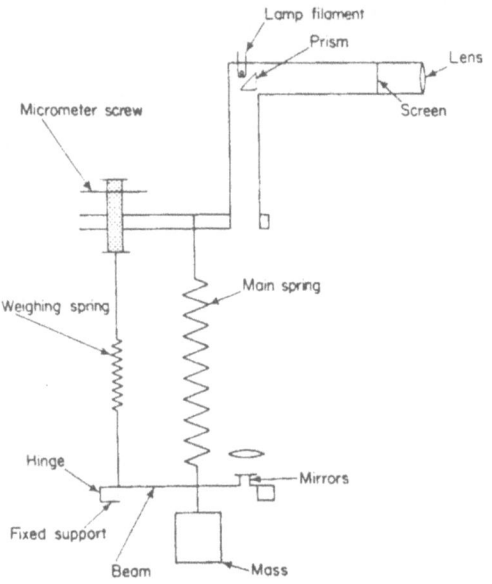

Figure 3.12. Hartley gravimeter

The two mirrors on the end of the beam are part of an optical lever. They are so mounted that when the beam moves vertically they rotate in opposite directions about horizontal axes thereby doubling the magnification. When the mirrors are coplanar the two images of the lamp filament are collinear. A light weighing spring is attached to the beam near the hinge and the tension in it is adjusted by means of a micrometer screw.

The instrument is used as follows. At a place where the value of g is known the micrometer screw is adjusted so that the two images of the filament are collinear. The instrument is taken to a second place where the value of g is also known and where it is slightly different from the value at the first station. The change in g causes a displacement of the two images and the micrometer screw is adjusted until the images are again collinear. The difference between the micrometer readings at the two places is a measure of the difference in g and hence the instrument is standardized. The effect of temperature is minimized by housing the entire instrument in an air-tight container kept at a constant temperature.

3.10 THE EÖTVÖS TORSION BALANCE

The Eötvös torsion balance is an extremely sensitive instrument which has been developed in order to measure very small variations in the value of g. These variations are produced by small geological deposits and the Eötvös balance is used extensively in surveying oil fields and for detecting heavy ores etc.

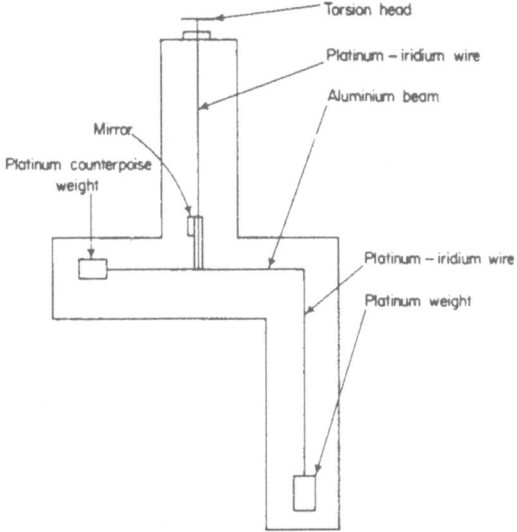

Figure 3.13. Eötvös torsion balance

The balance, shown diagrammatically in *Figure 3.13*, consists of a platinum–iridium torsion wire about 0·4 m long connected to a torsion head. The other end of the wire supports a horizontal aluminium beam 0·4 m in length. A 25 g cylindrical platinum weight is connected to one end of the beam by a platinum–iridium wire, 0·4 m long, and this is counterpoised by a platinum weight of about 30 g at the other end. A light aluminium rod is joined to the beam at its centre and this supports a small mirror which is used in

77

conjunction with a scale and telescope to measure the deflection of the beam.

If the value of g varies in the vicinity of the balance a couple acts on the suspended system and produces a twist in the wire. This causes the beam to be deflected from the position it would occupy if the value of g was constant. The formula relating the variation in g to the scale readings of the instrument may be derived as follows.

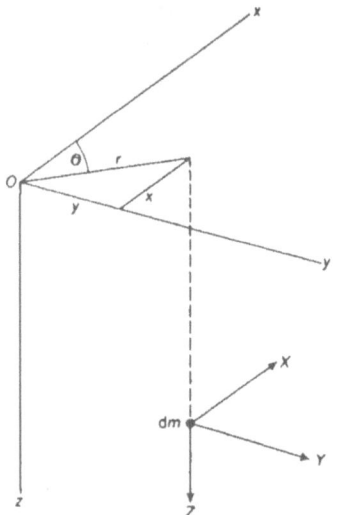

Figure 3.14. Eötvös torsion balance

When equilibrium is reached the beam of the balance is acted upon by gravity and by the torsion in the suspension. If U represents the gravity potential, let $\delta U/\delta x$, $\delta U/\delta y$ and $\delta U/\delta z$ be the values of the gravitational attraction along the north, east and vertical directions respectively. With reference to *Figure 3.14* the axes Ox, Oy are horizontal and Oz is vertical. If there is unit mass at the point (xyz) then the force acting on it along the axes are

$$\frac{\partial U}{\partial x} = X, \frac{\partial U}{\partial y} = Y \text{ and } \frac{\partial U}{\partial z} = Z$$

At the origin of coordinates, assuming the resultant force to be along Oz, i.e. $Z_0 = g_0$, then $X_0 = Y_0 = 0$. Now at the point (xyz)

$$X = X_0 + x\left(\frac{\partial X}{\partial x}\right)_0 + y\left(\frac{\partial X}{\partial y}\right)_0 + z\left(\frac{\partial X}{\partial z}\right)_0 + \ldots$$

$$Y = Y_0 + x\left(\frac{\partial Y}{\partial x}\right)_0 + y\left(\frac{\partial Y}{\partial y}\right)_0 + z\left(\frac{\partial Y}{\partial z}\right)_0 + \ldots$$

$$Z = Z_0 + x\left(\frac{\partial Z}{\partial x}\right)_0 + y\left(\frac{\partial Z}{\partial y}\right)_0 + z\left(\frac{\partial Z}{\partial z}\right)_0 + \ldots \quad \ldots (3.45)$$

If terms involving x^2, y^2, z^2, etc. and the second derivatives of X, Y and Z are neglected then

$$X = x\left(\frac{\partial^2 U}{\partial x^2}\right)_0 + y\left(\frac{\partial^2 U}{\partial x \partial y}\right)_0 + z\left(\frac{\partial^2 U}{\partial z \partial x}\right)_0$$

$$Y = x\left(\frac{\partial^2 U}{\partial x \partial y}\right)_0 + y\left(\frac{\partial^2 U}{\partial y^2}\right)_0 + z\left(\frac{\partial^2 U}{\partial y \partial z}\right)_0 \quad \ldots (3.46)$$

The torque about Oz is

$$\Gamma = \int (Yx - Xy)\, dm \quad \ldots (3.47)$$

where the integral is taken over the whole suspended system

$$\therefore \Gamma = \left(\frac{\partial^2 U}{\partial y^2} - \frac{\partial^2 U}{\partial x^2}\right)_0 \int xy\, dm + \left(\frac{\partial^2 U}{\partial x \partial y}\right)_0 \int (x^2 - y^2)\, dm$$

$$+ \left(\frac{\partial^2 U}{\partial y \partial z}\right)_0 \int zx\, dm - \left(\frac{\partial^2 U}{\partial z \partial x}\right)_0 \int yz\, dm \quad \ldots (3.48)$$

Let the beam be inclined at azimuth angle θ to the north–south direction in this equilibrium position. If r is the distance of the mass from z axis

$$x = r \cos\theta, \; y = r \sin\theta$$

$$\therefore \qquad \int xy\, dm = \frac{\sin 2\theta}{2} \int r^2 dm = \frac{I}{2} \sin 2\theta$$

$$\int (x^2 - y^2)\, dm = \cos 2\theta \int r^2 dm = I \cos 2\theta$$

where I is the moment of inertia about the z axis. Also, due to the symmetry about the axis of the beam

$$\int xz\, dm = \cos\theta \int rz\, dm = mhl \cos\theta$$

$$\int yz\, dm = \sin\theta \int rz\, dm = mhl \sin\theta$$

Hence, substituting in equation (3.48)

$$\Gamma = \frac{I}{2}\sin 2\theta \left(\frac{\partial^2 U}{\partial y^2} - \frac{\partial^2 U}{\partial x^2}\right)_0 + I\cos 2\theta \left(\frac{\partial^2 U}{\partial x \partial y}\right)_0$$

$$+ mhl\left(\frac{\partial^2 U}{\partial y \partial z}\cos\theta - \frac{\partial^2 U}{\partial z \partial x}\sin\theta\right) \quad \ldots (3.49)$$

At the equilibrium position

$$\Gamma = \tau\phi \qquad \ldots (3.50)$$

where ϕ is the angle of twist. If n_1 is the scale reading corresponding to the position when the beam makes angle θ with the north–south direction and n_0 the reading corresponding to a zero couple acting on the system, i.e. when $\Gamma = 0$, then

$$n_1 - n_0 = 2d \cdot \phi$$

where d is the distance from the mirror to the scale.
Hence

$$n_1 - n_0 = \frac{2d \cdot I}{\tau}\left[\left(\frac{\partial^2 U}{\partial y^2} - \frac{\partial^2 U}{\partial x^2}\right)_0 \frac{\sin 2\theta}{2} + \left(\frac{\partial^2 U}{\partial x \partial y}\right)_0 \cos 2\theta\right]$$

$$+ \frac{2d \cdot mhl}{\tau}\left[\left(\frac{\partial^2 U}{\partial y \partial z}\right)_0 \cos\theta - \left(\frac{\partial^2 U}{\partial z \partial x}\right)_0 \sin\theta\right] \quad \ldots (3.51)$$

Let

$$A = \frac{d \cdot I}{\tau} \text{ and } c = \frac{2dmhl}{\tau}, \text{ then}$$

$$n_1 - n_0 = A\sin 2\theta \left(\frac{\partial^2 U}{\partial y^2} - \frac{\partial^2 U}{\partial x^2}\right)_0 + 2A\cos 2\theta \left(\frac{\partial^2 U}{\partial x \partial y}\right)_0$$

$$- C\sin\theta \left(\frac{\partial^2 U}{\partial x \partial z}\right)_0 + C\cos\theta \left(\frac{\partial^2 U}{\partial y \partial z}\right)_0 \quad \ldots (3.52)$$

A and C are instrumental constants and can easily be determined. The value of n, at five different azimuthal angles is measured in order to determine n_0 and the terms, $(\partial^2 U/\partial y^2 - \partial^2 U/\partial x^2)_0$, $(\partial^2 U/\partial x \partial y)_0$, $(\partial^2 U/\partial x \partial z)$, and $(\partial^2 U/\partial y \partial z)_0$.

Thus the rate of change of g northwards, $\partial g/\partial x = \partial^2 U/\partial z \partial x$, and the rate of change eastwards, $\partial g/\partial y = \partial^2 U/\partial y \partial z$, can be determined.

The Eötvös balance is accurate to about 10^{-11} SI units compared with 10^{-6} SI units for an accurate pendulum experiment.

As mentioned previously, the balance is used in prospecting work and the values of the gravity gradients, $\partial g/\partial x$, $\partial g/\partial y$ provide important and useful information. The measurements are usually made in conjunction with electrical, magnetic, seismic and radioactive measurements and an experienced geophysicist can usefully interpret the results.

4

GRAVITATION

4.1 LAW OF GRAVITATION

THE law of gravitation may be said to have its beginning in the suggestion, put forward by Copernicus, that the planets went round the Sun. Accurate measurements on the positions of the planets were subsequently made in the fifteenth century by Tycho Brahe, and from these measurements Kepler was able to formulate his laws of planetary motion. At the same time Galileo was studying the laws of motion and reached the important conclusion that if a body is moving, then, if no forces act on the body, it will go on at a uniform speed in a straight line for ever. In order to explain the circular orbits of the planets Galileo postulated the existence of a force which acted on the planets deviating them from their natural path, i.e. a straight line, into a circular orbit. However, he did not ascribe this force to the attraction between the Sun and the planets.

Further relevant experiments were carried out in the sixteenth century by Gilbert, who put forward two more suggestions, first, that gravity is a reciprocal effect, i.e. if the Earth exerts an attractive force on a body, then the body exerts the same attraction on the Earth, and secondly, that gravity is not a property of a point in a body but of the whole matter of the body.

In the seventeenth century Newton, who had a better understanding of the laws of motion than the earlier workers, realized that the Sun could be the origin of the forces which governed the motion of the planets. He proved that since equal areas were swept out in equal times by a radius vector from the Sun to a planet, i.e. Kepler's second law, then all the forces acting on the planets are directed towards the Sun. An analysis of Kepler's third law, i.e. the squares of the periods of any two planets are proportional to the cubes of the major semi-axes of their respective elliptic orbits, showed Newton that the further away the planet, the weaker were the forces acting. Comparing two planets at different distances from the Sun, Newton was able to show that the forces acting were inversely proportional to the squares of the respective distances. Combining his conclusions, Newton said there must exist an attractive force between the planets and the Sun which varies inversely as the square of the distance between the planet and the Sun.

At this time it was known that the planet Jupiter had moons in orbit round it, just as the Earth's moon is in orbit round the Earth, and Newton suggested that the attractive force between the planets and their moons was of the same kind as that between the Sun and the planets.

Newton next examined the effect of the masses of the bodies on the attractive force between them. He knew from pendulum experiments that the Earth's gravitational force, exerted on a body, is proportional to the mass of the body. Considering the attraction to be a reciprocal process, then, the gravitational force is also proportional to the mass of the attracting body.

Newton then suggested that an attractive force exists between two bodies, and this is proportional to the masses of the two bodies concerned and inversely proportional to the square of the distance between them. He further proposed that this attractive force theory could be extended to all bodies and enunciated his universal law of gravitational attraction. This states that any mass m_1 attracts another mass m_2 at distance r away from it with a force F, along the line joining them, which is proportional to the product of the masses and inversely proportional to the square of their distance apart. It may be written as

$$F \propto \frac{m_1 m_2}{r^2}$$

or
$$F = G \cdot \frac{m_1 m_2}{r^2} \qquad \qquad \dots (4.1)$$

where G is the universal gravitational constant.

Newton tested the validity of this law in several ways. He tested the universality of the law by comparing the gravitational effect of the Earth on objects on its surface with the effect of the Earth on the moon. Calculations proved conclusively that bodies near the surface of the Earth are attracted by the same sort of force as is the moon. These calculations, completed in 1666, assumed that the whole mass of the Earth could be regarded as being at its centre. Newton did not prove that this assumption was valid until some years later, and hence his 'Universal Theory of Gravitation' was not published until 1687.

4.2 GRAVITATIONAL ATTRACTION AND POTENTIAL

According to the universal law of gravitational attraction, every body attracts every other body; this means that in the space surrounding each body, a gravitational field must exist. The intensity of the field at any point is defined as the force which would act on a particle of unit mass placed at that point, and this is a vector quantity. If the particle is moved in the gravitational field, then work must be expended. The amount of work may be positive or negative, according to the direction of movement of the particle. If a particle is moved from A to B in the gravitational field, then the work expended is equivalent to the difference in the gravitational potential between A and B. In order to obtain the value of potential at a particular point in the field it is usual to define zero potential as being at a point situated at infinity. The gravitational potential at any point in the gravitational field can then be defined as the work done per unit mass against the field in bringing up a small mass from infinity to that point.

If the field intensity at a point is F, then the work done in moving unit mass an infinitesimal distance ds at this point, against the attractive force is, force × distance or $F ds$. By the definition of potential, this work is equal to the difference in potential, dV, between the two points, hence

$$dV = F \, ds$$

i.e.
$$F = \frac{dV}{ds} \qquad \dots (4.2)$$

The difference in potential between two points in the gravitational field can then be found from the relationship

$$V = \int F \, ds \qquad \dots (4.3)$$

the limits of the integral being taken between the two points.

4.3 GRAVITATIONAL FIELDS

The gravitational fields due to certain types of bodies are of some importance and can be calculated as follows:

4.3.1 Thin spherical shell

With reference to *Figure 4.1*, O is the centre of a thin spherical shell of radius a. Let the surface density of the shell material be σ.

Divide the shell into thin circular rings with centres on OP, where P is a point at which the field strength is to be determined. The radius of one of these rings is thus $a \sin \theta$ and its thickness $a\,d\theta$. Every part of one of these rings is at the same distance from P, let this

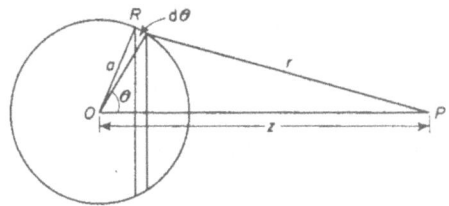

Figure 4.1. Gravitational field of a thin spherical shell

distance be r. Now, by symmetry, the direction of the field will be along OP. The field at P, along OP, due to one of the rings is given by

$$dF = G\,\frac{2\pi a \sin \theta \,.\, a\,d\theta \,.\, \sigma}{r^2}\cos RPO$$

i.e.
$$dF = G\,\frac{2\pi a^2 \sigma \sin \theta \,d\theta}{r^2}\,\frac{z - a\cos\theta}{r} \qquad \ldots\ (4.4)$$

where $OP = z$

From the diagram it is evident that

$$r^2 = z^2 + a^2 - 2za\cos\theta$$

Hence

$$r\,dr = za \sin \theta \,d\theta \qquad \ldots\ (4.5)$$

Also
$$z - a\cos\theta = \frac{z^2 + r^2 - a^2}{2z} \qquad \ldots\ (4.6)$$

Substituting equations (4.5) and (4.6) in (4.4)

$$dF = \frac{G\,2\pi\sigma a^2\,r\,dr}{r^2\,za}\,.\,\frac{z^2 + r^2 - a^2}{2zr} \qquad \ldots\ (4.7)$$

Hence
$$F = \frac{G\pi\sigma a}{z^2}\int_{z-a}^{z+a}\left(1 + \frac{z^2 - a^2}{r^2}\right)dr \qquad \ldots\ (4.8)$$

85

If P is outside the thin shell

$$F = \frac{G \cdot 4\pi a^2 \sigma}{z^2} \quad\quad \text{.... (4.9)}$$

but the mass of the sphere $M = 4\pi a^2 \sigma$ and thus

$$F = G \cdot \frac{M}{z^2} \quad\quad \text{...! (4.10)}$$

i.e. the gravitational field of the shell is as if the mass of the shell were concentrated at its centre O.

If P is on the surface of the shell

$$F = \frac{G\pi\sigma a}{z^2} \int_0^{2a} \left(1 + \frac{z^2 - a^2}{r^2}\right) dr \quad\quad \text{.... (4.11)}$$

To evaluate this integral it is necessary to assume P to be not quite on the sphere's surface so that $z = a + \delta$ where δ is very small. If this procedure is adopted then

$$F = \frac{G\pi\sigma a}{z^2} \int_\delta^{2a+\delta} \left(1 + \frac{z^2 - a^2}{r^2}\right) dr$$

$$F = \frac{G\pi\sigma a}{z^2} \int_\delta^{2a+\delta} \left(1 + \frac{2a\delta}{r^2}\right) dr$$

$$F = \frac{G 4\pi\sigma a^2}{z^2}$$

when δ becomes vanishingly small.

$$\therefore \quad\quad F = G \cdot \frac{M}{z^2} \quad\quad \text{.... (4.12)}$$

If P is inside the shell, then

$$F = \frac{G\pi\sigma a}{z^2} \int_{a-z}^{a+z} \left(1 + \frac{z^2 - a^2}{r^2}\right) dr \quad\quad \text{.... (4.13)}$$

$$\therefore \quad\quad F = 0$$

This shows that there is no resultant gravitational field inside the shell due to the shell itself.

4.3.2 Solid sphere or thick spherical shell

These can be considered as several concentric thin shells and the resultant gravitational field determined accordingly.

When P is outside the sphere the whole mass of the sphere acts as though it were concentrated at the centre O and hence over the range $OP = z = \infty$ to $z = a$

$$F = G \cdot \frac{M}{z^2} \qquad \qquad \ldots (4.15)$$

According to equation (4.14), if the point P is within the cavity of the thick shell, there is no resultant gravitational field.

When P is within the material of the sphere or thick shell, then the concentric thin shells external to the point will produce zero gravitational field but the shells inside will act as though all their mass were concentrated at the centre O. If the internal radius of the thick shell is assumed to be a_1 (for the solid sphere it is zero), then

$$F = \frac{G}{z^2} \frac{4}{3} \pi (z^3 - a_1^3) \rho \qquad \qquad \ldots (4.16)$$

where $\frac{4}{3}\pi(z^3 - a_1^3)\rho$ is the mass of material inside the point, ρ being the material density. In the case of a solid sphere, $a_1 = 0$, thus

$$F = \frac{G}{z^2} \frac{4}{3} \pi z^3 \rho$$

$$F = G \cdot \frac{M}{a^3} z \qquad \qquad \ldots (4.17)$$

i.e. the gravitational field which is directed towards the centre of the sphere is proportional to the distance of P from the centre.

4.4 GRAVITATIONAL POTENTIAL

Gravitational potential is also an important concept and it can be calculated according to equation (4.3). The potentials due to particular types of bodies are especially useful and can be calculated as follows.

4.4.1 A particle of mass M

The intensity of the gravitational field at a point, distance r from the particle, is GM/r^2 and is directed towards the particle.

Hence the work done in moving unit mass away from the particle through a small distance dr is $GM \, \mathrm{d}r/r^2$. Thus, the total work done in moving unit mass from a distance z to infinity is

$$- V_z = \int_z^\infty \frac{GM}{r^2} \, \mathrm{d}r = + \frac{GM}{z} \qquad \dots (4.18)$$

Hence the work done in bringing unit mass from infinity to the point, distance z from the particle is the potential at that point, thus

$$V_z = - \frac{GM}{z} \qquad \dots (4.19)$$

4.4.2 Thin spherical shell

With reference to *Figure 4.1*, the potential at the point P, distance z from O, due to a ring, distance r from P, is, according to equation (4.19),

$$\mathrm{d}V = - \frac{G \cdot 2\pi a \sin \theta \cdot a \, \mathrm{d}O \cdot \sigma}{r}$$

Thus from equation (4.5)

$$\mathrm{d}V = - \frac{G 2\pi a \sigma \, \mathrm{d}r}{z}$$

$$\therefore \qquad V_p = \int - \frac{G 2\pi a \sigma \, \mathrm{d}r}{z} \qquad \dots (4.20)$$

If P is outside the thin shell

$$V_P = - \frac{G 2\pi a \sigma}{z} \int_{z-a}^{z+a} \mathrm{d}r$$

$$\therefore \qquad V_P = - \frac{G 4\pi a^2 \sigma}{z}$$

$$= - \frac{GM}{z} \qquad \dots (4.21)$$

since $4\pi a^2 \sigma$ is the mass of the shell. Thus the potential at P is as if the whole mass of the shell were situated at its centre O.

If P is on the surface of the shell, $z = a$

$$\therefore \quad V_P = -G\frac{2\pi a\sigma}{z}\int_0^{2a} dr$$

$$= -\frac{GM}{z}$$

$$\therefore \quad V_P = -\frac{GM}{a} \qquad \qquad \dots (4.22)$$

If P is inside the shell

$$V_P = -G\frac{2\pi a\sigma}{z}\int_{a-z}^{a+z} dr$$

$$= G.4\pi a\sigma$$

$$\therefore \quad V_P = -\frac{GM}{a} \qquad \qquad \dots (4.23)$$

Thus, inside the shell, the potential is constant and equal to the potential at the surface of the shell.

4.4.3 Solid sphere or thick spherical shell

Again these configurations can be regarded as several concentric thin shells and the gravitational potentials determined accordingly.

When P is outside the sphere the mass of each constituent shell acts as though it were at the centre O, and thus the total potential at a point P is given by equation (4.21), i.e.

$$V_P = -\frac{GM}{z} \qquad \qquad \dots (4.24)$$

If P is within the cavity of the thick shell, then according to equation (4.23), a constituent shell of radius x and thickness dx, produces, at an internal point, a potential dV given by

$$dV = -\frac{G4\pi x^2\, dx\rho}{x}$$

where ρ is the density of the material of the thick shell. Hence, for a thick shell with internal and external radii equal to a_1 and a respectively

$$V_P = - G4\pi\rho \int_{a_1}^{a} x\, dx$$

$$\therefore \qquad V_P = - G2\pi\rho[a^2 - a_1^2] \qquad \qquad \dots (4.25)$$

But the mass (M) of the thick shell is given by

$$M = \tfrac{4}{3}\pi\rho[a^3 - a_1^3]$$

Hence

$$V_P = - \frac{3}{2} GM\left(\frac{a + a_1}{a^2 + aa_1 + a_1^2}\right) \qquad \dots (4.26)$$

If P is within the material of the sphere or thick shell such that $OP = z$, then the material exterior to P produces a potential given by equation (4.25), i.e.

$$V_{P1} = - G2\pi\rho(a^2 - z^2) \qquad \qquad \dots (4.27)$$

The material which is nearer to O than P, produces a potential at P given by equation 4.24, i.e.

$$V_{P2} = - \frac{G}{z}\frac{4}{3}\pi\rho(z^3 - a_1^3) \qquad \qquad \dots (4.28)$$

Thus the total potential at P is given by

$$V_P = V_{P1} + V_{P2} = - G2\pi\rho\left[a^2 - z^2 + \frac{2}{3z}(z^3 - a_1^3)\right] \qquad \dots (4.29)$$

$$= - \frac{2\pi G\rho}{3z}(-z^3 + 3za^2 - 2a_1^3)$$

$$V_P = - \frac{GM}{z}\left(\frac{3za^2 - 2a_1^3 - z^3}{2(a^3 - a_1^3)}\right) \qquad \dots (4.30)$$

In the case of a solid sphere, $a_1 = 0$ and substitution in equation (4.30) gives

$$V_P = - GM\left(\frac{3a^2 - z^2}{2a^3}\right) \qquad \qquad \dots (4.31)$$

4.5 MASS AND DENSITY OF THE EARTH

It has already been stated that the gravitational field of the Earth is an example of the universal law of gravitational attraction. Thus if the mass and radius of the Earth are denoted by M and R respectively, then, by equation (4.1), the gravitational force due to the Earth, exerted on a body of mass m on the Earth's surface is given by

$$F = G\frac{Mm}{R^2} \qquad \dots (4.32)$$

If the body is able to fall freely under the action of this force then its acceleration is given by F/m, which is the acceleration due to gravity, and hence,

$$g = \frac{GMm}{mR^2} = \frac{GM}{R^2} \qquad \dots (4.33)$$

The measurement of g has already been described. Since the value of the Earth's radius R is known from astronomical measurements, then a measurement of the gravitational constant G enables the mass of the Earth to be calculated or, vice versa, a measurement of M enables G to be calculated. If M is known, then assuming the Earth to be spherical, $M = 4\pi R^3 D/3$, where D is the mean density of the Earth. This density is that which a homogenous isotropic body of the same mass and volume of the Earth would have.

Early experiments attempted to measure M and thus to calculate G. These experiments are sometimes referred to as 'weighing the Earth', but this term is somewhat inaccurate since normally the expression 'weight' is used to refer to the attraction which the Earth exerts on a body near its surface and, by this definition, the Earth cannot be said to have any weight.

Later workers attempted to measure the gravitational constant G directly, by measuring the attraction between small masses in a laboratory. Many different experiments have been carried out and the more important are described below.

4.6 THE MEASUREMENT OF M
Basically, an experiment to measure M consists of comparing the attraction of a large terrestrial mass, i.e. a mountain, upon a plumb-line, with the attraction of the whole Earth on the same plumb-line. A knowledge of the local density of the mountain enables its mass to be calculated and hence the mass of the Earth may be determined.

Several experiments of this type were carried out but are now of historical interest only. The first experiments were performed by Bouguer in Peru in the mid eighteenth century. He measured the relative masses of a mountain and the Earth from the deflection of a plumb-line placed near the mountain. The mass of the mountain was estimated, inaccurately, from its volume and density. The value for the mass of the Earth was correspondingly inaccurate.

A few years later, Maskelyne, the Astronomer Royal, repeated the experiment in Scotland. He carried out an extensive survey of the hill, Schiehallion, in Perthshire, in order to obtain an accurate measurement of its mass and the position of its centre of gravity. Two stations were chosen at equal distances from the centre of gravity of the mountain and in a north-south line. With reference to *Figure 4.2*, suppose A and B represent the positions of the two stations either side of the mountain. AC and BD are the directions of the zeniths at these stations, which, when produced, intersect at the centre of the Earth, O. Let EA, EB be parallel rays from a distant fixed star, E. The zenith distances of the star are then α_1 and β_1 respectively. If the difference in geographical latitude between A and B is γ_1, then

$$\alpha_1 - \beta_1 = \gamma_1 \qquad \qquad \dots (4.34)$$

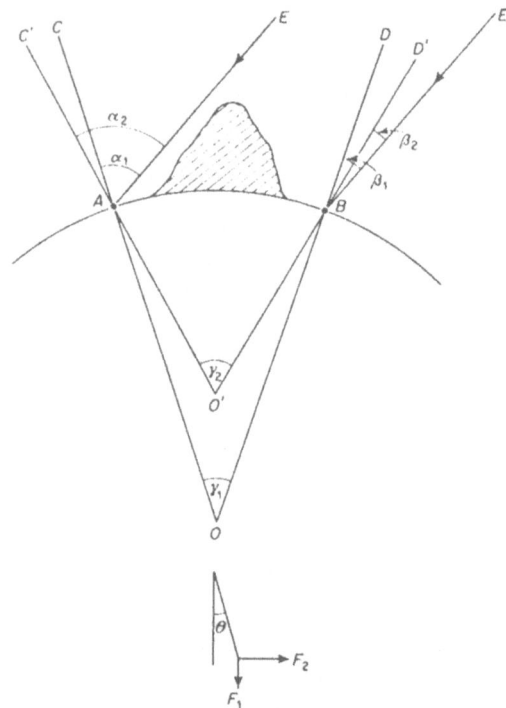

Figure 4.2. Maskelyne's experiment

However, due to the presence of the mountain, a plumb-line at A and B will not set along AC and BD but along AC' and BD' respectively, therefore these lines when produced intersect at O^1.

92

If the zenith distances as measured in the presence of the mountain are α_2 and β_2 respectively, then the difference in astronomical latitude between A and B is γ_2 where

$$\alpha_2 - \beta_2 = \gamma_2 \qquad \dots (4.35)$$

Hence the effect of the mountain is to make the astronomical latitude slightly greater than the geographical latitude, i.e. $\gamma_2 > \gamma_1$. Maskelyne's measurements, corrected for aberration, precession, etc. gave

$$\gamma_2 - \gamma_1 = 11\cdot6''$$

If the attraction of the Earth's field on the plumb-line is represented by F_1 and that of the mountain in a horizontal direction in the meridinal plane by F_2, then the deflection of the plumb-line from the vertical, θ, is given by

$$\tan \theta = \frac{F_2}{F_1} = \tan \frac{(\gamma_2 - \gamma_1)}{2} \qquad \dots (4.36)$$

Thus, in Maskelyne's measurements $\theta = 5\cdot8''$, F_2 was calculated from a knowledge of the mass of the mountain and the mean density of its constituents and the mass of the Earth was found. The value obtained was $5\cdot5 \times 10^{24}$ kg, which is not much below the best current value of $5\cdot98 \times 10^{24}$ kg. This type of experiment cannot afford very accurate results due to the difficulty in assessing the mass of the mountain. Nevertheless, both Bouguer's and Maskelyne's experiments are historically important.

Another type of experiment to determine M was carried out by Airy in 1854. The object of his experiment was to determine the mean density of the Earth by observing the difference in the periodic time of a pendulum at the top and bottom of a deep mine shaft. He used an invariable pendulum to determine the periodic times at the top and bottom of the shaft and these times were measured with great accuracy. The value of d was found to be $2\cdot5 \times 10^3$ kgm^{-3} by taking samples of the rock through which the shaft was bored. The experiment gave a value for D of $6\cdot5 \times 10^3$ kgm^{-3}. The currently accepted value of D is $5\cdot517 \times 10^3$ kgm^{-3}.

4.7 THE MEASUREMENT OF G

The first attempt to determine G by measuring directly the mutual attraction between two fairly small masses was performed by Cavendish towards the end of the eighteenth century. With reference to *Figure 4.3*, Cavendish suspended a light horizontal wooden rod, about 6 ft long by a silvered copper wire attached to a torsion head. At the ends of the rod were two small but heavy lead spheres of the same mass m_1. Two large lead spheres, of mass m_2 mounted on a frame, were placed near to the ends of the rod so that the centres of the four spheres were in a horizontal plane and each of the large spheres was near to one of the small spheres but on opposite sides

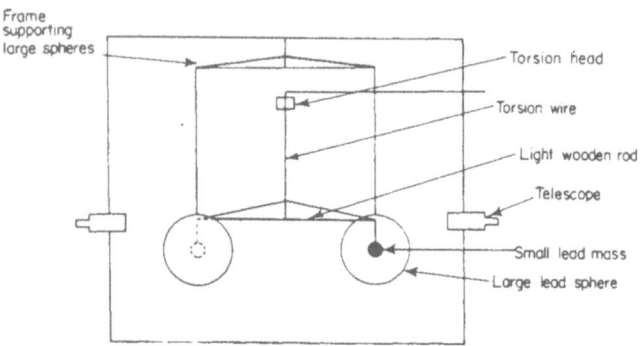

Figure 4.3. Determination of G by Cavendish

of the rod. The attraction between each small sphere and the adjacent large sphere thus produced a couple which caused the suspended system to move to a new equilibrium position where the restoring torque, due to the twist in the suspension wire was equal to the couple producing the displacement. Cavendish took many precautions in order to avoid temperature gradients and air currents which would cause erratic movements of the system. In spite of his precautions, however, the system was never in a position of static equilibrium and he had to estimate the equilibrium position by observing a number of consecutive swings of the rod.

Cavendish repeated the experiment several times and applied corrections for the attraction of each large sphere for the more distant small sphere which decreased the effective gravitational couple. Small corrections were also made for the effect of the frame supporting the large masses and for the attraction on the torsion beam, both of which tended to increase the gravitational couple. The mean result from his observations gave a value of 67.5×10^{-12} SI units for the gravitational constant G and hence a value of 5.45×10^3 kgm^{-3} for D.

Later experiments were carried out by various workers who made a number of improvements to the apparatus. An extremely accurate experiment was carried out by Boys in 1889 and this is described below.

4.8 BOYS' DETERMINATION OF G

Boys used quartz fibres instead of wire to support the masses. The torsional rigidity of quartz is much less than that of wire and hence a much greater deflection can be produced. Instead of observing the deflections by verniers moving on a scale, he used a mirror with an illuminated scale which further increased the sensitivity. These modifications permitted Boys to use much smaller equipment than Cavendish and consequently, air currents and temperature gradients, which had been impossible to avoid in the large room necessary to house the earlier equipment, were overcome. The main features of Boys' apparatus are shown diagrammatically in *Figure 4.4 (a)*.

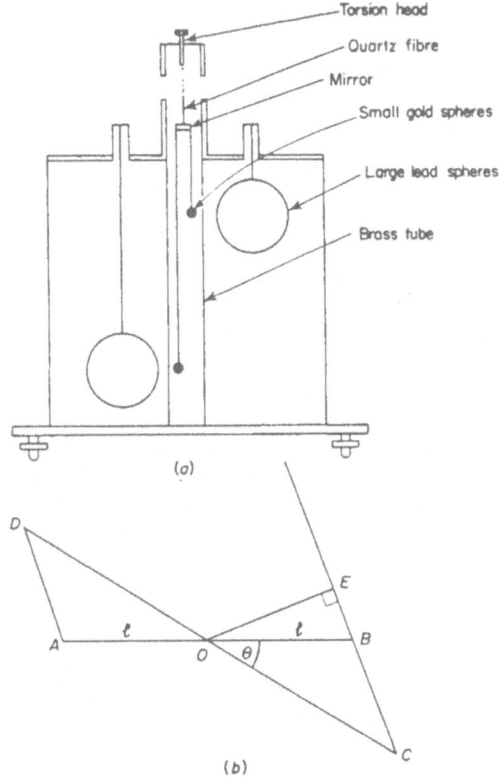

Figure 4.4. Determination of G by Boys

95

A quartz fibre about 0·4 m long connected to a torsion head supports a horizontal glass beam which acts as the mirror. Two other quartz fibres of an equal length, attached to the end of the mirror, support two small gold spheres of diameter less than 0·01 m. This suspension system is contained in a vertical cylindrical brass tube which protects the system from air currents, etc. Outside the brass tube two large lead spheres, about 0·1 m in diameter, are suspended at equal distances from the axis of the suspension system. The spheres are placed at heights so that they are situated in the same horizontal plane as the centres of the adjacent small gold spheres which they attract. The lead spheres are suspended from the lid of an outer brass case which can be turned about a vertical axis coincident with the axis of the central tube and thus the spheres can be arranged to deflect the beam first in one direction and then in the other as in the Cavendish experiment. Having the spheres at different levels considerably reduces the attraction of a lead sphere on the gold sphere at a different level.

The attraction between each gold sphere and the adjacent lead sphere produces a torque about the axis and causes the suspension system to be displaced through an angle ϕ to a position of equilibrium where the restoring torque due to the twist in the quartz fibre is equal to the gravitational torque. The position of the large lead spheres is chosen to make the moment of the gravitational torque as large as possible. After observing the deflection in one position the lead spheres are moved to the other side of the gold spheres and the change in deflection noted. This total displacement is equal to 2ϕ where ϕ is the angle of deflection of the suspension system from the zero equilibrium position.

With reference to *Figure 4.4* (*b*), suppose A and B represent the centres of the gold spheres, mass m_1, and C and D the centres of the lead spheres, mass m_2, when a position of equilibrium has been reached. The gravitational force along BC is Gm_1m_2/BC^2 and the moment of this force about O is $Gm_1m_2 . OE/BC^2$. Since there is an equal moment due to the force along AD, the total moment of the gravitational forces is

$$\frac{2Gm_1m_2}{BC^2} . OE$$

If the length of the beam $AB = 2l$, then $OE . BC = 2 .$ area $BOC = OB . OC \sin \theta = l . OC . \sin \theta$. The total moment of the gravitational forces is therefore

$$\frac{2Gm_1m_2l . OC \sin \theta}{BC^3} = \frac{2Gm_1m_2l . OC \sin \theta}{(OC^2 + l^2 - 2OC . l \cos \theta)^{\frac{3}{2}}} \quad \dots (4.37)$$

The restoring torque is $\tau\phi$ where τ is the torsional rigidity of the quartz suspension and ϕ is the measured angle of deflection of the suspension system from the zero position. At equilibrium therefore

$$\frac{2Gm_1m_2l \cdot OC \sin\theta}{(OC^2 + l^2 - 2OC \cdot l\cos\theta)^{\frac{3}{2}}} = \tau\phi \qquad \ldots (4.38)$$

and

$$G = \frac{\tau\phi(OC^2 + l^2 - 2OC \cdot l\cos\theta)^{\frac{3}{2}}}{2m_1m_2l \cdot OC \sin\theta} \qquad \ldots (4.39)$$

τ can be determined by measuring the period of small oscillations of the suspension system in the absence of the large masses, thus all the quantities in equation (4.39) can be determined and hence a value found for G.

Boys found a value of 66.5×10^{-12} SI units for G, and hence a value of $5.527 \times 10^3 \text{ kg m}^{-3}$ for D.

Braun performed similar experiments to Boys', but with somewhat larger apparatus. He carried out his experiments under a considerably reduced pressure and as a result air currents were much reduced. However, the most accurate value of G obtained so far is probably due to Heyl.

4.9 DETERMINATION OF G BY HEYL

Heyl's apparatus is shown in *Figure 4.5 (a)*. His suspension system consists of a tungsten wire about 1 m in length connected to an aluminum rod about 0.2 m long, from the ends of which small spheres are suspended with their centres in the same horizontal plane. The experiment was performed with spheres of different materials, gold, platinum and optical glass being used in three experiments. The spheres are relatively heavy so that the moment of inertia of the suspended system is nearly all due to the spheres. For large attracting masses large steel cylinders are used whose mass is about 66 kg. These are placed so that their centres of gravity are in the same horizontal plane as the centre of gravity of the small spheres. By having all the centres of gravity in the same horizontal plane it is possible to measure the distances between the masses with considerable accuracy.

The large steel cylinders are suspended from a framework which can be rotated about a vertical axis coincident with the axis of the suspension system. This enables the cylinders to be positioned so that their centres are in line with the centres of the spheres as shown

GRAVITATION

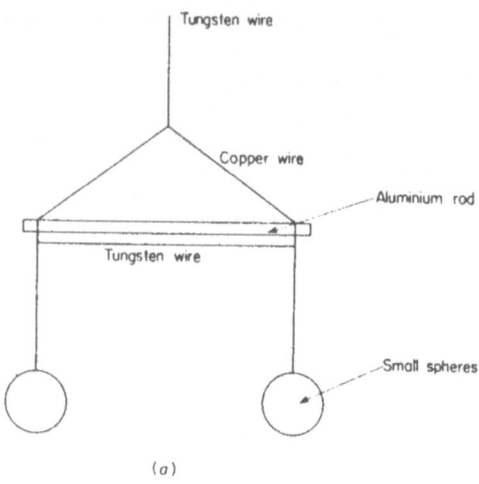

(a)

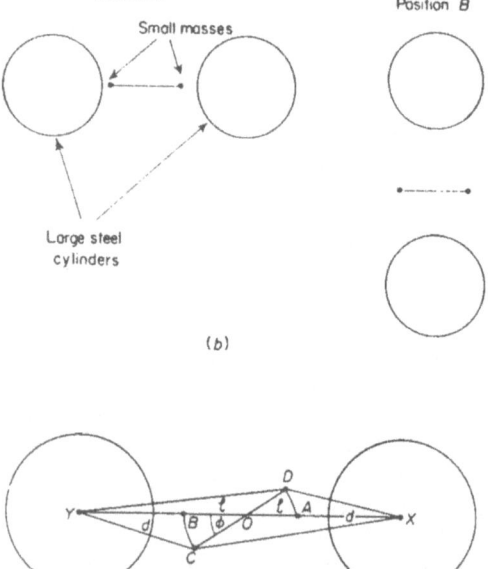

(c)

Figure 4.5. Determination of G by Heyl

98

in *Figure 4.5* (*b*) position *A*, or so that the line between their centres cuts the line between the centres of the sphere at right angles as in position *B*. The suspension system is enclosed in a brass container and the air pressure reduced to a few millimetres of mercury, so as to reduce air currents, etc. The two spheres are set into oscillation with a small amplitude of a few degrees and the periodic time measured with the large cylinders, first, in the *A* position and then in the *B* position. In the *A* position the gravitational attraction between the spheres and cylinders accelerates the swing, while in the *B* position the swing is retarded.

When the cylinders are in the *A* position the theory of the experiment is as follows. With reference to *Figure 4.5* (*c*), suppose *X* and *Y* are the centres of the large cylinders, *A* and *B* are the centres of the spheres when at rest, and *C* and *D* the centres of the spheres when displaced through an angle ϕ The energy of the system consists of (*a*) kinetic energy equal to $I\dot{\phi}^2/2$, where I is the moment of inertia of the suspension system about the vertical axis of rotation; (*b*) potential energy due to torsion in the wire equal to $K\phi^2/2$, where K is the torsional constant of the wire and is not assumed to be known; (*c*) gravitational potential energy whose magnitude can be calculated when the spheres are at the positions *C* and *D*. The gravitational potential energy of the displaced suspension system can be written as $GA_1\phi^2/2$, where A_1 is the product of a large and a small mass and is a geometrical function of $OA = l$ and $AX = d$.

Since total energy is constant (neglecting friction),

$$\frac{I\dot{\phi}^2}{2} + \frac{K\phi^2}{2} + \frac{GA_1\phi^2}{2} = \text{constant} \qquad \ldots (4.40)$$

Differentiating with respect to time and simplifying

$$I\ddot{\phi} + (K + GA_1)\phi = 0 \qquad \ldots (4.41)$$

This represents simple harmonic motion and the period of oscillation is given by

$$T_1 = 2\pi\sqrt{\frac{I}{(K + GA_1)}} \qquad \ldots (4.42)$$

When the cylinders are in the *B* position the equation of motion is

$$I\ddot{\phi} + (K + GA_2)\phi = 0 \qquad \ldots (4.43)$$

where A_2 is the new geometrical constant in this position. The period of oscillation is thus

$$T_2 = 2\pi \sqrt{\frac{I}{(K + GA_2)}} \qquad \dots (4.44)$$

Eliminating the torsional constant K from equations (4.42) and (4.44)

$$G = \frac{4\pi^2 I(T_2^2 - T_1^2)}{(A_1 - A_2) T_1^2 T_2^2} \qquad \dots (4.45)$$

The periodic times can be measured very accurately, and since I, A_1 and A_2 can all be calculated accurately, the method provides an extremely accurate method for the determination of G.

In Heyl's experiments the gold spheres were found to absorb mercury from the air in the laboratory, thereby increasing their mass. The results with gold spheres were therefore rejected. Slightly different values of G were obtained with the platinum and glass spheres, which is not explicable. The mean value of G obtained by Heyl was 66.7×10^{-12} SI units. Further measurements carried out by Heyl and Chrzanowski in 1942 gave the same result.

4.10 DETERMINATION OF G BY POYNTING

In 1881 von Jolly described an experiment to measure G which involved the use of a common balance. The principle of his experiment was to counterpoise a balance and then to disturb the balance by placing under the mass in one of the scale pans a large mass. Due to the gravitational attraction between the masses the balance was disturbed, and to restore the balance to its original position an additional mass was added to the other scale pan. From a measurement of the quantities involved G could be calculated. von Jolly performed such an experiment but, due to the very small deflections involved and the other factors such as convection currents, his result was not very accurate.

The experiment was repeated in 1891 by Poynting who eliminated many of the inaccuracies in von Jolly's work. His apparatus is represented diagrammatically in *Figure 4.6*. A large bullion balance was used with a steel beam about 1·2 m long. Equal spheres, m_1, weighing about 20 kg and made of a lead–antimony alloy were suspended from each end of the beam. The large mass m_2, placed under one of these spheres in order to disturb the balance was a sphere weighing about 150 kg, also made of the lead–antimony alloy. This mass was mounted on a special table which could be rotated about a vertical axis so that the mass could be brought vertically under either of the spheres suspended from the balance

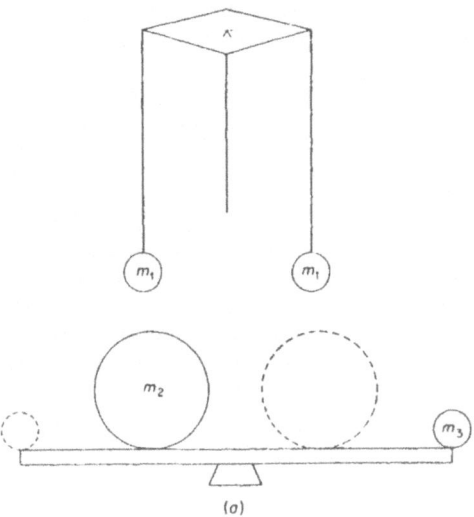

(a)

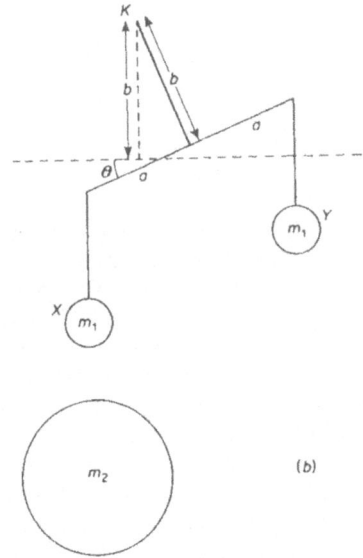

(b)

Figure 4.6. Determination of G by Poynting

101

E

beam. In order to avoid tilting the floor as the large mass was moved it was necessary to use a compensating mass m_3 on the table. This mass was about twice the distance of m_2 from the axis of rotation of the table and its mass was about half that of m_2. The attraction between the large mass m_2 and the sphere immediately above it is equal to Gm_1m_2/d^2, where d is the distance between the centres of the masses. This force results in the balance beam being deflected downwards on the side where the mass is placed. If the mass is rotated on its table so as to be under the other sphere m_1, then the balance beam will be deflected downwards on to the other side.

The deflection produced is very small and, in order to measure it accurately, the Kelvin double-suspension mirror is used. This is illustrated in *Figure 4.7* and consists of a mirror supported by

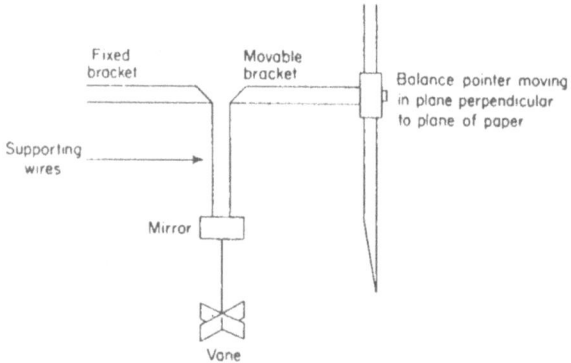

Figure 4.7. Kelvin double-suspension mirror

two wires, one of which is attached to a fixed bracket and the other to a movable bracket. The movable bracket is attached to the pointer of the balance at right angles. The pointer moves in a plane perpendicular to that of the diagram. Thus when the balance beam tilts slightly, the pointer moves slightly, thereby causing the wire attached to the movable bracket to move. This causes the mirror to rotate about the stationary wire attached to the fixed bracket. By the use of this device Poynting succeeded in magnifying the deflection of the balance beam by a factor of 150. The effect of air currents on this very sensitive system is minimized by the use of damping vanes which are immersed in oil.

If it is assumed that the balance beam is horizontal when the large mass m_2 is not present, the equilibrium of the balance when it

102

is tilted through an angle θ due to the mass m_2 being placed under one of the spheres, say X, as in *Figure 4.6*, is represented by the equation

$$\left(m_1 g + \frac{Gm_1 m_2}{d^2}\right)(a \cos \theta - b \sin \theta) = m_1 g(a \cos \theta + b \sin \theta)$$

$$\ldots (4.46)$$

where $2a$ is the length of the balance beam and b is the distance of its midpoint from the knife edge.

If the large mass m_2 is removed and a small extra mass is attached to the sphere X in order to cause the balance beam to be tilted through the same angle θ again, then

$$(m_1 g + m'g)(a \cos \theta - b \sin \theta) = m_1 g(a \cos \theta + b \sin \theta) \ldots (4.47)$$

Combining equations (4.46) and (4.47)

$$\frac{Gm_1 m_2}{d^2} = m'g \qquad \ldots (4.48)$$

i.e.

$$G = \frac{m'gd^2}{m_1 m_2} \qquad \ldots (4.49)$$

The large mass m_2 also exerts an attraction on the balance beam itself and this can be corrected for as follows. The experiment is repeated with the two spheres X and Y several centimetres above m_2 (as in the first experiment). Thus the attraction of m_2 on X and Y is changed, while the attraction on the balance beam, from the first experiment, is unchanged. Equation (4.48) for the two experiments then becomes

$$\frac{Gm_1 m_2}{d_1^2} + Z = m'g \qquad \ldots (4.50)$$

and

$$\frac{Gm_1 m_2}{d_2^2} + Z = m''g \qquad \ldots (4.51)$$

where Z is the force between m_2 and the beam. Eliminating Z from equation (4.50) and (4.51) gives

$$G = (m'' - m')g/m_1 m_2(1/d_2^2 - 1/d_1^2) \qquad \ldots (4.52)$$

Corrections also have to be applied for the gravitational attraction between m_2 and the sphere Y, not directly above it, as well as for the gravitational attraction between the compensating mass m_3 and the spheres X and Y.

Poynting carried out his experiments with the balance placed in an underground room and totally enclosed it, so as to reduce air currents, temperature gradients, etc. All adjustments were made by remote control mechanisms from outside the room and the deflection of the mirror was observed through a telescope. The small extra weight m' added to the sphere m_1, in order to produce the same deflection θ as the effect of the gravitational attraction, consisted of a small rider of mass about 10 mg which was moved a measured distance along the balance beam.

Poynting obtained a value of G of 66.98×10^{-12} SI units.

4.11 VARIATION OF G

Many experiments have been carried out by various workers to detect changes which may occur in the value of the universal gravitational constant when conditions are varied. Early work by Eötvös with the torsion balance showed that G remained constant within the range of experimental error when a wide variety of materials were used as the attracting masses. His experimental error was very small and the results showed that G was constant to within $10^{-9}G$.

The effect of the temperature of the masses was studied by Shaw over the range 0°C to 250°C and again no variation in G was observed.

Several workers, notably Eötvös, Austin and Thwing studied the effect of the intervening medium between the attracting masses. Slabs of various materials were interposed between the pairs of attracting spheres in apparatus similar to that of Boys, but no change in G could be detected.

Recent work by Dicke has confirmed these earlier results and it must be concluded that if there is any variation in G, then it is extremely small.

5

ELASTICITY

5.1 ELASTICITY

IN a previous chapter the motion of a rigid body was discussed.
It was assumed at that stage that a rigid body is one in which the
distance between any two points in it is invariable no matter what
external forces may act on the body. In practice no body is completely rigid and the subject of elasticity is concerned with the
behaviour of materials under the action of external forces. Since
no body is perfectly rigid, then a body, when subjected to an external
force, undergoes a change of shape or size or both and it is said to
be in a state of strain. If the body regains completely its original
shape and size when the external force or stress is removed the
body is said to be perfectly elastic. On the other hand, if the body
completely retains its altered shape and size when the external
force is removed then the body is termed perfectly plastic. These
terms have a very limited application, however, since materials
which are perfectly elastic only regain their original shape and size
if the external force is not excessively large. It is also probable that
even a perfectly plastic material recovers partially if it is acted
upon by only a small external force.

5.2 STRESS AND STRAIN

The external force or forces constituting the stress exerted on a
body can be applied in various ways and thus the type of strain
produced varies with the nature of the force. Probably the simplest
type of stress is that due to a force applied along the longitudinal
axis of a rod giving rise to an increase in length of the rod. In such
a case the strain is measured by the relative increase in length of the
rod, i.e. change in length per unit length.

Alternatively, the body may be subjected to a change in pressure,
i.e. uniformly compressed in all directions, and this type of stress
gives rise to a volume strain which is measured as the change in
volume per unit volume.

A third type of stress is a shear stress. This type of stress is
illustrated in *Figure 5.1*, where it can be seen that the unit cube

of material is subjected to the action of the tangential forces F. For equilibrium, a shear stress in a given direction cannot exist without an equal shear stress at right angles to it and these forces are also shown on the diagram. If the cross-section of the cube is initially represented by $ABCD$ then, after the application of the shear stress, the face $ABCD$ will be displaced to $A'B'CD$ where the

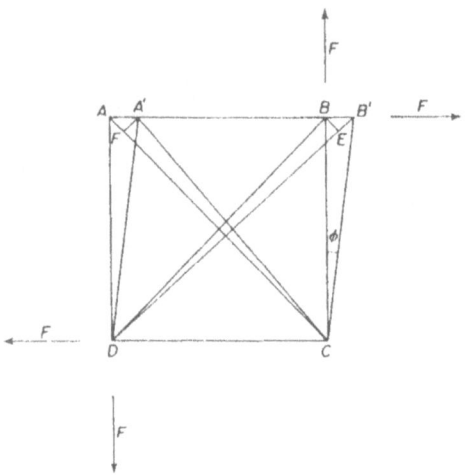

Figure 5.1. Shear strain

angle $ADA' = \phi$. The shear strain produced is measured by the angular deformation ϕ, and this is equal to the relative lateral displacement of two horizontal layers in the material at unit distance apart, i.e. with reference to the figure, for small values of ϕ

$$\phi = \tan \phi = \frac{AA'}{AD} = \frac{\text{relative displacement}}{\text{distance apart}} \quad \dots (5.1)$$

It is also possible to regard a shear strain as a combination of an extension and a contraction at right angles to the extension. Consider *Figure 5.1*, it can be seen that the diagonal DB is increased in length to DB' while the diagonal AC is shortened to $A'C$. If BE is drawn perpendicular to DB', then, since BB' is small, $BB'E = 45°$ and $B'EB = 90°$, and thus, $BE = B'E$. The strain along DB is therefore

$$\frac{DB' - DB}{DB} = \frac{BE}{DB} = \frac{BB'}{\sqrt{2}} \cdot \frac{1}{\sqrt{2BC}} = \frac{BB'}{2BC} = \frac{\phi}{2} \quad \dots (5.2)$$

106

Similarly, it can be shown that the contraction along the diagonal *AC* is $\phi/2$. Hence the shear ϕ is equivalent to an extension $\phi/2$ together with a contraction of $\phi/2$ at right angles to each other and at 45° to the direction of shear.

5.3 HOOKE'S LAW

For most materials it has been established experimentally that over a considerable range the strain is proportional to the stress. If the stress is increased to a point where this relationship is no longer valid then the elastic limit of the material is said to have been exceeded.

The linear relationship between strain and stress up to the elastic limit was discovered by Hooke in the seventeenth century and is known as Hooke's law. Provided the material is subjected to a stress below the elastic limit it recovers its original shape and size when the stress is removed. If the applied stress exceeds the elastic limit, however, the material does not regain its original shape and size when the stress is removed and it is said to acquire a permanent set.

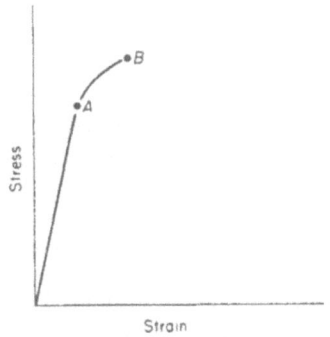

Figure 5.2. Stress–strain graph of cast iron

The elastic limit of a material can also be defined as the stress which produces the maximum amount of recoverable deformation. It may be determined experimentally by subjecting the material to a stress and by measuring the dimensions of the body upon removal of the stress. The process is repeated while the stress is gradually increased until it is found that the body does not regain its original shape and size completely after removal of the stress, i.e. the body has acquired a permanent set. A graph, *Figure 5.2*, of the strain against the applied stress shows where the permanent set begins.

107

This type of graph results when a brittle material such as cast iron is subjected to stresses. Initially, a straight line graph is obtained, which is the region where Hooke's law is obeyed. At the point A, the elastic limit of the material is reached. The graph is seen to curve over until finally the material breaks at the point B.

5.4 ELASTIC MODULI

Provided the elastic limit of a material is not exceeded then, according to Hooke's law, the ratio of stress to strain is constant and the value of this constant is known as the elastic modulus of the material. Since the stress may be applied in various ways, giving rise to different types of strain, a material possesses a number of elastic moduli.

In the case of a rod subjected to a force applied along its longitudinal axis giving rise to an increase in the length of the rod, the ratio of stress to strain is known as Young's modulus and is usually given the symbol E. If the applied force is F and the radius of the rod of circular cross-section is r, then the stress is equal to $F/\pi r^2$. The strain is equal to dl/l where dl is the increase in length of the rod initially of length l. Thus

$$E = \frac{\text{stress}}{\text{strain}} = \frac{F}{\pi r^2} \cdot \frac{l}{dl} \qquad \dots (5.3)$$

For a laboratory experiment where the force is produced by hanging weights on the rod, $F = mg$. An experiment to determine Young's modulus for a material in the form of a long thin wire is described later.

If a body is subjected to a shear stress, then, provided the elastic limit is not exceeded, the ratio of shear stress to shear strain is a constant and is known as the rigidity modulus of the material. The shear stress p is the tangential force per unit area and if this produces a shear strain ϕ, then

$$n = p/\phi \qquad \dots (5.4)$$

where n is the rigidity modulus.

If a body is subjected to a change in pressure, giving rise to a volume strain, then the ratio of stress to strain is known as the bulk modulus of the material and is usually represented by the symbol K. If the initial volume v of the body increases to a volume $v + \delta v$ when the pressure is increased from p to $p + \delta p$, then, since an increase in pressure gives rise to a decrease in the volume, the resulting strain for the stress δp is $-\delta v/v$.

Hence $$K = \frac{\delta p}{(-\delta v/v)} = -v\left(\frac{\partial p}{\partial v}\right) \qquad \ldots\ldots (5.5)$$

Since p is a function of other variables such as temperature, the partial differential coefficient is necessary. Frequently the compressibility of a material is referred to and this quantity is the reciprocal of the bulk modulus of the material.

A fourth elastic modulus is occasionally referred to which is given by the ratio of stress to strain when a rod is extended in length without any change in its lateral dimensions. In the case of a simple extension then the rod obviously contracts laterally. In order to produce an extension without lateral contraction of the rod the stress producing the extension must be accompanied by two perpendicular stresses of sufficient magnitude to prevent lateral contraction. In this case the ratio of the extensional stress to the longitudinal strain is known as the axial modulus, χ of the material.

A straightforward longitudinal extension produced by a suitable stress acting along the longitudinal axis, is, as mentioned above, accompanied by a lateral contraction. In the case of a stretched wire the ratio of the decrease in diameter of the wire to its initial diameter is termed the lateral strain. Provided the elastic limit of the material is not exceeded it is found that the ratio of the lateral strain to the longitudinal strain is constant and this constant is known as Poisson's ratio, σ, for the material.

The various elastic moduli defined above are not independent constants but are all interrelated. This is easy to understand since it is possible to produce any type of strain by a suitable combination of both volume and shear strains. The relationship between the elastic moduli is discussed in the following section.

5.5 RELATION BETWEEN THE ELASTIC MODULI

Consider a unit cube of material to be subjected to a volume stress of magnitude P. Since the bulk modulus K is defined as stress/strain, then the volume strain is equal to P/K. If a perpendicular stress P only acted on one pair of opposite faces of the cube, e.g. along the Ox axis, the extensional strain normal to those forces would be equal to P/E by definition of Young's modulus. At the same time the cube would undergo a contraction equal to $\sigma P/E$ along the Oy and Oz axes by definition of Poisson's ratio. Similarly, application of the stress P on the other two pairs of opposite faces at the cube causes an extensional strain normal to the faces of magnitude P/E and a contraction of $\sigma P/E$ along the axes at right

109

E *

angles to the direction of the stress. *Table 5.1* illustrates clearly the effect of applying such stresses. In this table the first columns indicate the stresses along the axes, while the second group of columns indicate the resulting strains. For example, $a + P$ in the Ox column simply means that a stress is applied in the Ox direction so as to extend the cube. The result is an extension of P/E in the Ox direction and a contraction of $\sigma P/E$ in the Oy and Oz directions. From this table it is clear that the total extensional strain ε along each of the axes is

$$\frac{P}{E} - \frac{2\sigma P}{E} = \frac{P}{E}(1 - 2\sigma) = \varepsilon$$

Now the total volume strain is equal to the increase in volume divided by the initial volume

i.e. $$\frac{(1 + \varepsilon)^3 - 1^3}{1^3} = +3\varepsilon = \frac{3P}{E}(1 - 2\sigma) \qquad \dots (5.6)$$

neglecting higher powers of ε.

Since the volume strain is equal to P/K, then

$$\frac{P}{K} = \frac{3P}{E}(1 - 2\sigma) \qquad \dots (5.7)$$

or $$E = 3K(1 - 2\sigma) \qquad \dots (5.8)$$

Table 5.2 shows the effect of applying a perpendicular stress to produce an extension along the Ox axis and a perpendicular stress to produce a contraction along the Oy axis. Combination of these two stresses results in an extensional strain of magnitude $P/E(1 + \delta)$ along Ox and a contraction of $P/E(1 + \sigma)$ along Oy. Now it has previously been shown that a shear strain ϕ is equivalent to an

Table 5.1

Stress applied			Strain produced		
Ox	Oy	Oz	Ox	Oy	Oz
$+P$	0	0	$+P/E$	$-\sigma P/E$	$-\sigma P/E$
0	$+P$	0	$-\sigma P/E$	$+P/E$	$-\sigma P/E$
0	0	$+P$	$-\sigma P/E$	$-\sigma P/E$	$+P$
$+P$	$+P$	$+P$	$P/E(1 - 2\sigma)$	$P/E(1 - 2\sigma)$	$P/E(1 - 2\sigma)$ Sum

Table 5.2

Stress applied			Strain produced		
Ox	Oy	Oz	Ox	Oy	Oz
$+P$	0	0	$+P/E$	$-\sigma P/E$	$-\sigma P/E$
0	$-P$	0	$+\sigma P/E$	$-P/E$	$+\sigma P/E$
$+P$	$-P$	0	$P/E(1+\sigma)$	$-P/E(1+\sigma)$	0 Sum

extension $\phi/2$ together with a compression $\phi/2$ at right angles to each other and both at 45° to the direction of shear. Hence the strains in this case are equivalent to a shear strain of magnitude $2P/E(1+\sigma)$ at 45° to Ox or Oy.

Now the equivalent shearing stress is equal to P and by the definition of rigidity modulus, the shear strain is equal to P/n. Thus

$$\frac{P}{n} = \frac{2P}{E}(1+\sigma) \qquad \dots (5.9)$$

or

$$E = 2n(1+\sigma) \qquad \dots (5.10)$$

From equations (5.8) and (5.10) it is clear that if any two of the four quantities, E, K, n and σ, are known, the others may be calculated.

The elimination of E from equations (5.8) and (5.10) gives Poisson's ratio in terms of the bulk modulus and rigidity modulus, thus

$$\sigma = \frac{3K - 2n}{6K + 2n} \qquad \dots (5.11)$$

Since the rigidity modulus is essentially a positive quantity the minimum value of n is zero, therefore the maximum value of Poisson's ratio is 0·5. The minimum value of σ must be a positive quantity since a negative value of σ would mean that a material extended in one direction would also expand in a direction normal to the direction in which it is being stretched and no material behaves in this manner. Hence

$$0 < \sigma < 0·5 \qquad \dots (5.12)$$

and for many materials σ is approximately 0·33.

5.6 TORSION OF A CYLINDER

With reference to *Figure 5.3a*, let the cylindrical rod be fixed at the end O and twisted at the other end through an angle ϕ by an external

111

couple of moment Γ, where the axis is that of the cylinder, OO'. Let the length of the cylinder be l and its radius a. The cylinder is in a state of pure shear since there can be no change in either the length or radius of the cylinder.

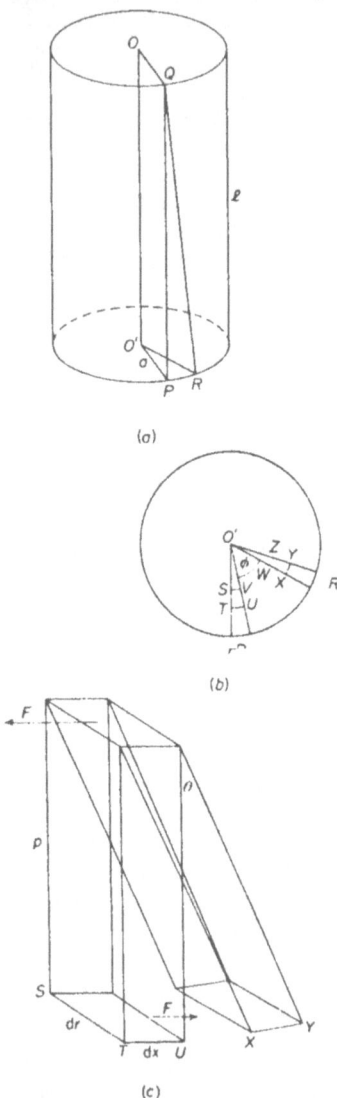

(a)

(b)

(c)

Figure 5.3. Torsion of a cylinder

The angular displacement ϕ is proportional to the magnitude of the applied couple Γ and to the distance from the fixed end of the cylinder l. The action of the couple causes the cylinder to be twisted so that a wedge of the cylinder such as $OO'PQ$ is twisted into the position $OO'RQ$ where angle $PO'R$ is equal to ϕ. *Figure 5.3b* shows a cross-section through the end of the cylinder which is twisted through the angle ϕ and a typical element of this cross-section $STUV$ is twisted through ϕ to the position $WXYZ$. *Figure 5.3c* shows how the section of the cylinder with $STUV$ as its base is thus displaced to the new position where the angle of shear is θ.

Let this movement be produced by a force F acting tangentially on $STUV$ and let $O'S = r$, $ST = dr$ and $TU = dx$. Thus, the shear stress acting on $STUV$ is of magnitude $F/dr\,dx$. Since this produces a shear strain of θ, then

$$n = \frac{F}{\theta \cdot dr \cdot dx} \qquad \ldots \text{(5.12)}$$

where n is the rigidity modulus of the material.

Now $l\theta = r\phi$, and hence

$$F = n \cdot dr \cdot dx \cdot \frac{r\phi}{l} \qquad \ldots \text{(5.13)}$$

The moment of the force about the axis OO' of the cylinder is Fr and the total moment may be obtained by integration, hence

$$\Gamma = \frac{n\phi}{l} \iint r^2 dr\,dx \qquad \ldots \text{(5.14)}$$

where the integral of dx is taken round the circle of radius r and the integral of r is taken between the limits of O and a, for a solid cylinder

$$\Gamma = \frac{n\phi}{l} \cdot 2\pi \int_0^a r^3\,dr \qquad \ldots \text{(5.15)}$$

$$\therefore \qquad \Gamma = \frac{\pi n a^4}{2l}\phi \qquad \ldots \text{(5.16)}$$

If the cylinder is hollow with outer radius equal to a and inner radius equal to a_1 then the integral is taken between these limits to give

$$\Gamma = \frac{\pi n \phi}{2}(a^4 - a_1^4) \qquad \ldots \text{(5.17)}$$

113

The torque required to produce unit twist between the ends of the rod, i.e. Γ / ϕ is known as the torsional rigidity and is usually given the symbol τ. Equation (5.16) affords a convenient method of determining the rigidity modulus for a cylindrical specimen since n can easily be evaluated by measuring the angular deformation produced by a known applied couple. Suitable experiments based on the application of this formula are described later.

The energy in a rod of circular cross-section under torsion may be calculated as follows. When the applied couple is Γ, the work done in twisting the rod through an angle $d\phi$ is

$$dV = \Gamma \, d\phi \qquad \qquad \dots (5.18)$$

Hence, from equation (5.16)

$$dV = \frac{\pi n a^{4}}{2l} \phi \, d\phi \qquad \qquad \dots (5.19)$$

Consequently

$$V = \frac{\pi n a^4}{2l} \int_{0}^{\phi} \phi \, d\phi$$

$$V = \frac{\pi n a^4 \phi^2}{2l \cdot 2}$$

$$\therefore \qquad V = \frac{\Gamma \phi}{2} \qquad \qquad \dots (5.20)$$

5.7 BENDING OF A BEAM

If a beam is acted on by an external couple, the beam is bent so that the filaments of the beam in the region nearer the inside of the curve are compressed, while the filaments which are in the region nearer the outside of the curve are extended. One particular filament remains unchanged in length when the beam is bent and this is called the neutral filament or neutral axis.

Figure 5.4 illustrates part of a beam *ABCD* bent so as to form part of a circle with radius R. A filament such as *GH*, distance z from the neutral filament *EF* has a length equal to $(R + z)\phi$ where ϕ is the angle *EF* subtends at the centre of curvature. The extension e of the filament is given by

$$e = GH - EF = (R + z)\phi - R\phi = z\phi \qquad \dots (5.21)$$

The original length of the filament was $R\phi$ and thus the strain is equal to $z\phi/R\phi = z/R$. If the cross-sectional area of the filament is α and if p is the force acting across the area to produce the extension, then $$\frac{p}{\alpha} = E\,\frac{z}{R}\qquad\qquad \dots(5.22)$$

i.e. $$p = E\,\frac{z\alpha}{R}\qquad\qquad \dots(5.23)$$

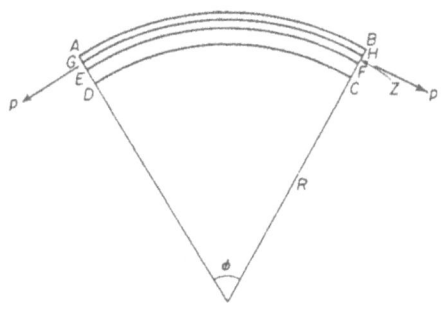

Figure 5.4. Bent beam

Forces such as p are produced above and below the neutral filament and these forces constitute a system of couples whose resultant is termed the moment of resistance or the internal bending moment of the beam. The moment of p about the neutral filament is pz and the internal bending moment is

$$\sum pz = \frac{E}{R}\sum \alpha z^2 \qquad\qquad \dots(5.24)$$

When the beam is in equilibrium the internal bending moment must be equal to the external applied couple. $\sum \alpha z^2$ is a quantity analogous to the moment of inertia about the neutral filament and is called the geometrical moment of inertia about that axis. If the cross-sectional area of the beam is A and the radius of gyration is k then $\sum \alpha z^2 = Ak^2$. Hence

$$\text{Internal bending moment} = \frac{EAk^2}{R} \qquad\qquad \dots(5.25)$$

5.8 ENERGY IN A BENT BEAM

Let the length of the neutral filament EF in *Figure 5.4* be dl. The work done in stretching the filament GH a distance $e = z\phi$ is given by

115

Force × distance = Stress × area of cross-section × distance

= Elastic modulus × strain × area × distance

$$\therefore \qquad = \int_0^e E\alpha \frac{e}{dl} \cdot de$$

$$= \frac{1}{2} E\alpha \frac{e^2}{dl} \qquad \qquad \dots (5.26)$$

But $e = z\phi$ and $dl = R\phi$

$$\therefore \qquad e = z\frac{dl}{R}$$

Hence,

$$\text{work done} = \frac{1}{2}\frac{E\alpha z^2}{R^2} dl$$

The total energy of the whole cross-section A of the beam of length dl is

$$dV = \frac{1}{2}\frac{E}{R^2} dl \sum \alpha z^2$$

$$dV = \frac{dl}{2R^2} EAk^2 \qquad \qquad \dots (5.27)$$

The energy of the whole beam is thus

$$V = \int_0^l \frac{EAk^2}{2R^2} \cdot dl \qquad \qquad \dots (5.28)$$

5.9 THE CANTILEVER

The cantilever consists of a uniform horizontal rod fixed at one end into a rigid support. *Figure 5.5* illustrates such a cantilever OB fixed at O and bent by the application of a load at the free end B. At any point in the cantilever such as C, the internal forces acting at the point C must keep the part CB of the beam in equilibrium together with the applied force W. Now the force W acting vertically downwards at B is balanced by an equal vertical force acting upwards at C and together these forces constitute a couple of moment $W \times CB$. This is called the bending moment at C. Equilibrium of the cantilever is maintained by a bending moment of equal magnitude but of opposite sense, i.e. the internal bending moment. The internal forces acting in the cantilever thus consist of the internal bending moment whose magnitude is EAk^2/R according to equation (5.25)

116

together with a shearing stress of magnitude W/ab where a is the width and b the depth of the cantilever.

The internal and 'external bending moments at the point C may be equated as follows. With axes Ox and Oy along, and perpendicular to, the unstrained horizontal position of the beam, let

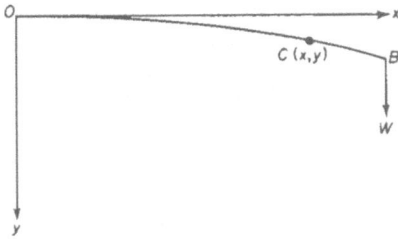

Figure 5.5. The cantilever

the coordinates of the point C be (x, y) and let the length of the cantilever $AB = l$. The external bending moment at C is $W(l - x)$. Hence

$$W(l - x) = \frac{EAk^2}{R} \qquad \dots (5.29)$$

Now the radius of curvature at C is given by the expression

$$\pm \frac{1}{R} = \frac{d^2y/dx^2}{\left(1 + \left(\frac{dy}{dx}\right)^2\right)^{\frac{3}{2}}}$$

If the curvature of the cantilever is small $(dy/dx)^2 \ll 1$, and thus

$$\frac{1}{R} \simeq \pm \frac{d^2y}{dx^2}$$

Substituting in equation (5.29)

$$W(l - x) = EAk^2 \frac{d^2y}{dx^2} \qquad \dots (5.30)$$

Integrating,

$$EAk^2 \frac{dy}{dx} = W\left(lx - \frac{x^2}{2}\right) + C_1$$

117

When $x = 0$, $\dfrac{dy}{dx} = 0$, therefore $C_1 = 0$

and hence

$$EAk^2 \frac{dy}{dx} = W\left(lx - \frac{x^2}{2}\right) \qquad \dots (5.31)$$

Integrating,

$$EAk^2 y = W\left(\frac{lx^2}{2} - \frac{x^3}{6}\right) + C_2$$

When $x = 0$, $y = 0$ and therefore $C_2 = 0$
Hence,

$$EAk^2 y = W\left(\frac{lx^2}{2} - \frac{x^3}{6}\right) \qquad \dots (5.32)$$

The maximum depression from the horizontal position at B is given by substituting $x = l$ in equation (5.32). Thus

$$y_{max} = \frac{Wl^3}{3EAk^2} \qquad \dots (5.33)$$

For a cantilever of rectangular cross-section $k^2 = b^2/12$, and since $A = ab$

$$y_{max} = \frac{4Wl^3}{Eab^3} \qquad \dots (5.34)$$

The shearing stress W/ab causes a shear of the cantilever and thus also causes a lowering of the end B. This is very small, however, and can normally be neglected. If the rigidity modulus of the cantilever material is n, then the depression of the end B due to the shear stress is given by

$$y_{shear} = \frac{Wl}{abn} \qquad \dots (5.35)$$

Now

$$\frac{y_{shear}}{y_{max}} = \frac{Wl}{abn} \cdot \frac{Eab^3}{4Wl^3} = \frac{E}{4n}\left(\frac{b}{l}\right)^2 \qquad \dots (5.36)$$

For a long thin cantilever $(b/l)^2$ is extremely small and y_{shear} is negligible compared with y_{max}.

The simple treatment above applies only to the case of a light cantilever i.e. one whose own weight may be neglected. If the cantilever has appreciable weight this must be taken into account. Let the weight per unit length of the cantilever be w. The external bending moment at C is then increased by a force of magnitude $w(l - x)$ acting at the centre of gravity of CB, i.e. at $(l - x)/2$ from C. Hence, equating the internal and external bending moments

$$EAk^2 \frac{d^2 y}{dx^2} = W(l - x) + \frac{w}{2}(l - x)^2 \qquad \ldots (5.37)$$

Integrating

$$EAk^2 \frac{dy}{dx} = W\left(lx - \frac{x^2}{2}\right) + \frac{w}{2}\left(l^2 x - lx^2 + \frac{x^3}{3}\right) + C_1$$

Since $\frac{dy}{dx} = 0$ when $x = 0$ then $C_1 = 0$. On integrating again,

$$EAk^2 y = W\left(\frac{lx^2}{2} - \frac{x^3}{6}\right) + \frac{w}{2}\left(\frac{l^2 x^2}{2} - \frac{lx^3}{3} + \frac{x^4}{12}\right) + C_2$$

Since $y = 0$ when $x = 0$, $C_2 = 0$ and hence

$$EAk^2 y = W\left(\frac{lx^2}{2} - \frac{x^3}{6}\right) + \frac{w}{2}\left(\frac{l^2 x^2}{2} - \frac{lx^3}{3} + \frac{x^4}{12}\right) \ldots (5.38)$$

The maximum depression from the horizontal position at B is thus given by the equation

$$EAk^2 y_{max} = \frac{Wl^3}{3} + \frac{wl^4}{8} = \frac{l^3}{3}\left(W + \frac{3W_1}{8}\right)$$

where $W_1 = wl$ is the total weight of the cantilever.

5.10 BEAM LOADED AT ITS CENTRE AND SUPPORTED NEAR ITS ENDS
Figure 5.6 shows a light beam AB supported near its ends being subjected to an applied load W at its centre O. There is thus, a force of magnitude $W/2$ acting vertically at each support. The section of the beam at O is horizontal and each half of the beam may be regarded as being equivalent to a cantilever, firmly fixed at O and subjected to a force $W/2$ at its end where that end is distance $l/2$ from O assuming $AB = l$. In this case the end of each cantilever is raised and not depressed from the horizontal position. The maximum elevation of A and B from the horizontal position is given by equation (5.33) where l is equal to $l/2$ and $W = W/2$ in this case.

Hence the distance of A or B above O is given by the equation

$$y = \frac{Wl^3}{48EAk^2} \qquad \dots (5.40)$$

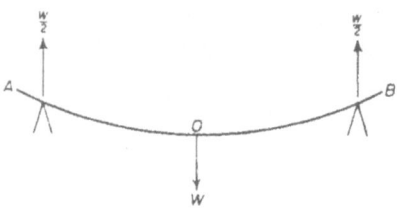

Figure 5.6. Beam loaded at its centre

Young's modulus for the material of a beam can be determined, using equation (5.33) or (5.40). Experimental details are provided later.

5.11 BENDING OF A BEAM UNDER A THRUST

Figure 5.7 illustrates a vertical beam AB loaded in the direction of its length, W being the load applied at each end of the beam. The beam is assumed to bend with a small radius of curvature. If the maximum horizontal displacement of the beam from the vertical is a, then with axes Ox and Oy as shown, the internal bending moment and external bending moment at the point C are related by the equation

$$EAk^2 \cdot \frac{d^2y}{dx^2} = W(a - y) \qquad \dots (5.41)$$

The solution of equation (5.41) is

$$y = a\left(1 - \cos\sqrt{\frac{W}{EAk^2}} \cdot x\right) \qquad \dots (5.42)$$

If the length of the beam is l, then when $x = l/2$, $y = a$ and hence

$$\cos\sqrt{\frac{W}{EAk^2}} \cdot \frac{l}{2} = 0$$

$$\therefore \qquad W = \frac{\pi^2 EAk^2}{l^2} \qquad \dots (5.43)$$

This equation is independent of a and gives the critical value of W below which no bending occurs and the beam is in stable equilibrium.

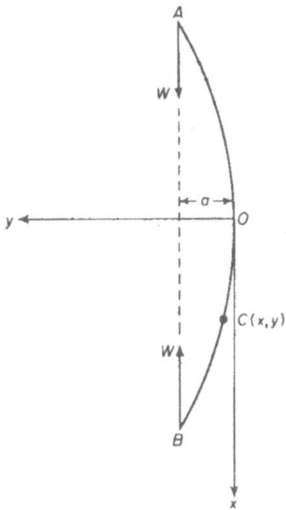

Figure 5.7. Loading of a vertical beam

If the beam is rigidly clamped at each end then its shape under the applied load is somewhat different, as shown in *Figure 5.8*. The length *DE* of the beam is under similar conditions of bending

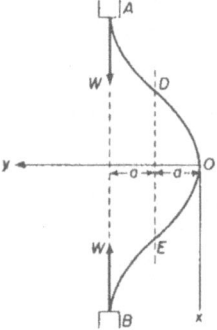

Figure 5.8. Rigidly clamped vertical beam

as before and, in this case, $y = a$ when $x = l/4$ and the critical load is therefore given by the equation

$$W = \frac{4\pi^2 E A k^2}{l^2} \qquad \qquad \dots (5.44)$$

121

5.11.1 Example

A uniform cantilever of weight W and length l is clamped horizontally at one end. What vertical force must be applied at the free end to raise it to the level of the clamped end?

Let the weight per unit length of the cantilever be w. Hence $W = wl$. If P is the applied vertical force at the free end then with reference to *Figure 5.9*, at the point C, equating the internal and external bending moments

$$E A k^2 \frac{d^2 y}{dx^2} = \frac{w}{2}(l - x)^2 - P(l - x)$$

Integrating

$$E A k^2 \frac{dy}{dx} = \frac{w}{2}\left(l^2 x - l x^2 + \frac{x^3}{3}\right) - P\left(lx - \frac{x^2}{2}\right) + C_1$$

When $x = 0$, $dy/dx = 0$, and hence $C_1 = 0$. Integrating

$$E A k^2 y = \frac{w}{2}\left(\frac{l^2 x^2}{2} - \frac{l x^3}{3} + \frac{x^4}{12}\right) - P\left(\frac{l x^2}{2} - \frac{x^3}{6}\right) + C_2$$

When $x = 0$, $y = 0$, and hence $C_2 = 0$
When $x = l$, let $y = \delta$

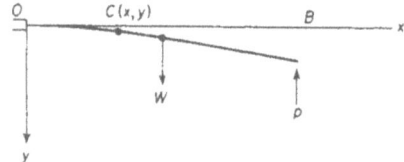

Figure 5.9. Example

Thus
$$E A k^2 . \delta = \frac{w}{2}\left(\frac{l^4}{2} - \frac{l^4}{3} + \frac{l^4}{12}\right) - P\left(\frac{l^3}{2} - \frac{l^3}{6}\right)$$

$$= \frac{W l^3}{8} - \frac{P l^3}{3}$$

Hence for δ to be zero

122

$$\frac{Wl^3}{8} = \frac{Pl^3}{3} \text{ or } P = \frac{3W}{8}$$

5.12 DETERMINATION OF YOUNG'S MODULUS FOR A LONG THIN WIRE

In the next few sections several well-known methods for determining the elastic constants of different materials available in various forms are described.

For a material available in the form of a long thin wire, Young's modulus can be directly determined by loading the wire with various known masses and measuring the extension produced. *Figure 5.10* illustrates the experimental arrangement. Two wires of

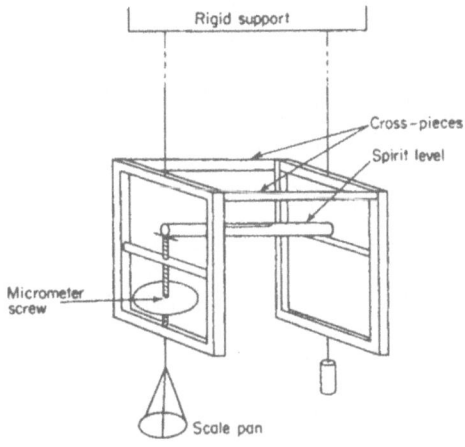

Figure 5.10. Determination of Young's modulus for a wire

the material are suspended from a rigid support parallel to each other. The lower ends of the wires are connected to a frame which is provided with a sensitive spirit level. The mass of the framework is sufficient to cause the wires to be in a stretched state. The cross-pieces shown on the diagram enable the frames attached to each wire to move vertically with respect to each other but prevent relative rotation of the frames. The spirit level can be adjusted by the micrometer screw thus enabling the vertical movement to be measured. Weights are added to the scale pan shown, extending one of the wires with respect to the other. Since the two wires are affected equally by temperature changes during the experiment, a temperature correction is not necessary. After the addition of each weight to the scale pan the micrometer screw is adjusted until the spirit level is horizontal and the resulting vertical movement measured. Loading of the wire is continued and measurements obtained of the extension for each load. Care must be taken to avoid exceeding the elastic limit of the material under test. The wire is then gradually unloaded and again readings of the extension are noted. The mean value of the extension for each load is tabulated and a graph plotted of the load against the extension. The original length of the wire under test is carefully measured and the radius of the wire determined by a micrometer screw gauge. At least six readings at the radius at various places along the wire are necessary to give an average value. Young's modulus may then be calculated by substitution in equation (5.3).

Theoretically it is possible to determine Poisson's ratio for the material in the experiment described above by measuring the decrease in the radius at the wire produced and thus calculating the lateral strain. In practice, for a thin wire the lateral strain is so small that it cannot be measured with any degree of accuracy, but in industry, Poisson's ratio is determined by direct measurement of longitudinal and lateral strains for materials in the form of a thick rod. In such an experiment, very large forces must be applied to produce appreciable longitudinal extensions.

5.13 DETERMINATION OF POISSON'S RATIO FOR INDIA-RUBBER

A long solid cylindrical piece of india-rubber is supported at one end from a rigid support and a scale pan is attached to the other end. The diameter of the rubber is of the order of 1–2 cm. The tube is marked at several places along its length and the diameter at each place determined by a micrometer screw gauge. Weights are gradually added to the scale pan and after allowing a steady equilibrium state to be reached after the addition of each weight, the extension of the rubber is noted and the diameters at the places marked are measured. The mean lateral contraction of the rubber is calculated from the latter readings.

Poisson's ratio may be simply calculated from the results according to the formula

$$\sigma = \frac{\text{lateral contractions}}{\text{original diameter}} \times \frac{\text{original length}}{\text{longitudinal extension}}$$

5.14 DETERMINATION OF YOUNG'S MODULUS FOR THE MATERIAL OF A BEAM

In Section 5.10 it was shown that if a light beam is supported on knife edges near its ends and a load is applied at the centre of the beam then the maximum depression of the centre of the beam below the horizontal position is given by the equation

$$y_{max} = \frac{Wl^3}{48EAk^2}$$

Hence Young's modulus is given by

$$E = \frac{Wl^3}{48Ak^2 y_{max}} = \frac{mgl^3}{4ab^3 y_{max}} \qquad \dots (5.45)$$

where mg is the applied load, l is the distance between the knife edges, a is the width and b the depth of the beam.

The experiment is carried out by supporting a beam about 1 m long on two knife edges near its ends, the distance between the knife edges being accurately measured. A pan is suspended from a knife edge which is placed on the beam at its mid-point and the load is applied by placing masses in the pan. The depression of the centre of the beam is conveniently determined by the use of a vernier scale, one part of which is fixed, the other moving with the beam. The readings of maximum depression, y_{max}, against applied load, m, are recorded and a graph plotted, m/y_{max} being determined from the slope. The width and depth of the beam are measured and this must be done accurately since b is raised to the third power in the formula for E. Since all the quantities in the formula are known E may then be determined.

5.15 DETERMINATION OF YOUNG'S MODULUS OF THE MATERIAL OF A BEAM BY THE VIBRATION METHOD

In the case of a cantilever, weight W_1 firmly clamped at one end and loaded at the other with a weight W, it was shown that the maximum depression at the free end is given by equation (5.39)

$$EAk^2 y_{max} = \frac{Wl^3}{3} + \frac{W_1 l^3}{8}$$

If a further displacement y is made by applying a further force F at the free end, then

$$EAk^2(y_{max} + y) = l^3 \left(\frac{W + F}{3} + \frac{W_1}{8} \right)$$

and hence

$$F = \frac{3EAk^2 y}{l^3} \qquad \dots (5.46)$$

125

The force F is balanced by the internal stresses in the beam, and the potential energy associated with these stresses is given by

$$\int_0^y F \, dy = \frac{3}{2} \frac{EAk^2}{l^3} y^2 \qquad \ldots (5.47)$$

The potential energy of the cantilever with reference to its equilibrium position under the action of the weight W as a standard is thus

$$\frac{3}{2} \frac{EAk^2}{l^3} y^2 - Wy - W_1 u \qquad \ldots (5.48)$$

where u is the additional distance by which the centre of the beam has been depressed. The equilibrium position of the centre of the beam under the action of the weight W alone, u_0, can be calculated from equation (5.38) by putting $y = u_0$ when $x = l/2$. Thus

$$EAk^2 u_0 = l^3 \left(\frac{5W}{48} + \frac{17W_1}{384} \right) \qquad \ldots (5.49)$$

Since u denotes the additional displacement resulting from the increase in applied load of F

$$u = \frac{5Fl^3}{48EAk^2} \qquad \ldots (5.50)$$

Hence from equation (5.46)

$$u = \frac{5}{16} y \qquad \ldots (5.51)$$

The equation for potential energy can thus be expressed as

$$\text{P.E.} = \frac{3}{2} \frac{EAk^2}{l^3} y^2 - Wy - \frac{5W_1 y}{16} \qquad \ldots (5.52)$$

If the cantilever is released by removing the extra force F there is an upward force tending to restore the weight W to its original equilibrium position and the cantilever vibrates.

Now equation (5.38) gives the total displacement, y, at any point with coordinate x. When W is increased by F, y increases by an amount s where, from equation (5.46)

$$s = \frac{F}{EAk^2} \left(\frac{lx^2}{2} - \frac{x^3}{6} \right) = \frac{3}{l^3} \left(\frac{lx^2}{2} - \frac{x^3}{6} \right) y$$

Hence the velocity of the beam at any point is given by

$$\dot{s} = \frac{3}{l^3}\left(\frac{lx^2}{2} - \frac{x^3}{6}\right)\dot{y}$$

and the kinetic energy of an element of the beam, length dx, is given by

$$\frac{1}{2}\frac{w}{g}\,dx\left[\frac{3}{l^3}\left(\frac{lx^2}{2} - \frac{x^3}{6}\right)\dot{y}\right]^2 \qquad \text{.... (5.53)}$$

where w/g is the mass per unit length of the beam.

The total kinetic energy of the whole beam is obtained by integrating between 0 and l for x. This gives

$$\text{Total kinetic energy} = \frac{33}{280}\frac{w}{g}l\dot{y}^2 = \frac{33}{280}\frac{W_1}{g}\dot{y}^2 \quad \text{.... (5.54)}$$

Also, the kinetic energy of the load at the end of the beam is

$$\frac{1}{2}\frac{W\dot{y}^2}{g} \qquad \text{.... (5.55)}$$

The total energy of the beam and load is the sum of potential energy and kinetic energy, i.e.

$$\frac{1}{2}\left(\frac{W}{g} + \frac{33}{140}\frac{W_1}{g}\right)\dot{y}^2 + \frac{3}{2}\frac{EAk^2}{l^3}y^2 - Wy - \frac{5W_1y}{16} \quad \text{.... (5.56)}$$

Since the total energy is constant, on differentiating equation (5.56) and dividing by $\dot{y}$, there results

$$\left(\frac{W}{g} + \frac{33}{140}\frac{W_1}{g}\right)\ddot{y} + 3\frac{EAk^2}{l^3}y - \left(W + \frac{5W_1}{16}\right) = 0 \quad \text{.... (5.57)}$$

If
$$q = y - \frac{l^3\left(W + \dfrac{5W_1}{16}\right)}{3EAk^2}$$

Then, equation (5.57) becomes

$$\left(\frac{W}{g} + \frac{33}{140}\frac{W_1}{g}\right)\ddot{q} + \frac{3EAk^2}{l^3}q = 0 \qquad \text{.... (5.58)}$$

127

This equation represents simple harmonic motion with period

$$T = 2\pi \sqrt{\frac{\left(\dfrac{W}{g} + \dfrac{33}{140}\dfrac{W_1}{g}\right)l^3}{3EAk^2}} \qquad \dots (5.59)$$

This formula can then be used to determine E since W, W_1, l and Ak^2 can be determined prior to the experiment and T can be measured accurately.

The beam is clamped rigidly to a solid support at one end and a mass attached to the free end. The mass should not be so great as to produce an excessive depression. The beam is set into vibration and T carefully measured. The experiment can be repeated for various lengths of the vibrating beam and for various masses suspended from the end. When the length of the beam is varied care must be taken to calculate the weight of the beam W_1 which refers only to the weight of the vibrating part of the beam. If T^2 is plotted against W/g for a number of experiments carried out at a fixed value of l, then the graph has a slope of $4\pi^2 l^3 / 3EAk^2$ and the negative intercept is of magnitude, $33W_1/140g$.

5.16 DETERMINATION OF THE RIGIDITY MODULUS OF A WIRE BY A STATICAL METHOD

Equation (5.16) derived in Section 5.6 may be applied directly in order to determine the rigidity modulus of a wire. In an experiment described by Barton, the specimen in the form of a uniform cylindrical wire is suspended vertically from a rigid support as shown in *Figure 5.11*. The other end of the specimen is firmly attached to a solid brass cylinder B. A torque is applied to the wire by placing weights in the scale pans. Flexible cords attached to the scale pan pass over the pulleys and act tangentially on the brass cylinder. The angle of twist ϕ between two points on the specimen, distance l apart is measured by attaching two mirrors to the specimen at the points and using the usual lamp and scale. If the weight in each pan is mg and the radius of the cylinder is r, then the applied couple is $2mgr$. Therefore, according to equation (5.16)

$$2mgr = \frac{\pi n a^4 \phi}{2l} \qquad \dots (5.60)$$

The radius of the wire must be determined accurately by means of a

128

micrometer screw gauge at various points along the wire since it occurs to the fourth power in the equation.

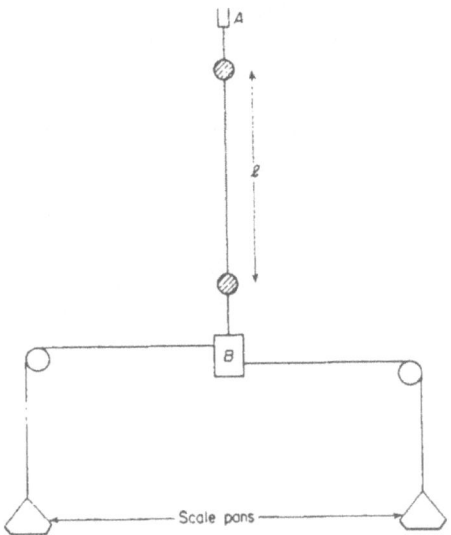

Figure 5.11. Determination of rigidity modulus by static method

5.17 DETERMINATION OF THE RIGIDITY MODULUS OF A WIRE BY A DYNAMICAL METHOD. MAXWELL'S NEEDLE

If a horizontal bar is supported at its mid point from the end of a vertical wire, then a small displacement from its equilibrium position causes the bar to perform simple harmonic oscillations about this position. If the bar is displaced through an angle ϕ the stresses produce a restoring couple of magnitude $\tau\phi$ and, if the moment of inertia of the bar about the vertical axis of rotation is I, the equation of motion is

$$I\ddot{\phi} + \tau\phi = 0 \qquad \ldots\ldots (5.61)$$

So that the period of oscillation is given by

$$T = 2\pi\sqrt{\frac{I}{\tau}} \qquad \ldots\ldots (5.62)$$

Hence, if I is known, T can be measured and τ determined. If τ is known then the modulus of rigidity of the wire can be calculated since $\tau = \pi n a^4/2l$.

129

In Maxwell's needle the 'bar' consists of a hollow tube of length d. Four equal cylinders each of length $d/4$ fit into the tube, two being solid brass and two being hollow. Let I_0 be the moment of inertia of the hollow cylinder of length d about the wire as axis, I_1 be the moment of inertia of the solid brass cylinder about a parallel axis through its centre of gravity, I_2 be the moment of inertia of the hollow brass cylinder about a parallel axis through its centre of gravity, m_1 the mass of each solid brass cylinder and m_2 the mass of each hollow brass cylinder.

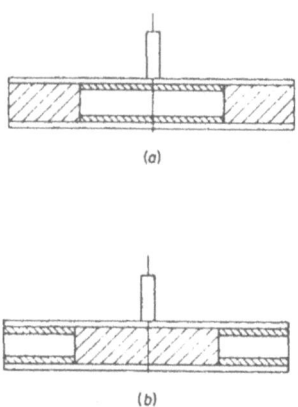

(a)

(b)

Figure 5.12. Maxwell's needle

The cylinders are placed in the hollow tube with the hollow cylinders in the centre as shown in Figure 5.12(a). The tube is slightly displaced from its equilibrium position and released so that it oscillates about this position. The periodic time of the oscillations, T_1 is determined. A second experiment is then performed with the cylinders arranged as in Figure 5.12(b), i.e. with the solid cylinders in the centre of the tube, and the periodic time T_2 measured. If the moment of inertia of the complex bar in the first experiment is I' and in the second experiment is I'', then

$$T_1 = 2\pi\sqrt{\frac{I'}{\tau}} \text{ and } T_2 = 2\pi\sqrt{\frac{I''}{\tau}}$$

thus

$$T_1^2 - T_2^2 = \frac{4\pi^2}{\tau}(I' - I'') \qquad \ldots\ldots(5.63)$$

Now, by application of the theorem of parallel axes

$$I' = I_0 + 2I_2 + 2m_2\left(\frac{d}{8}\right)^2 + 2I_1 + 2m_1\left(\frac{3d}{8}\right)^2$$

$$I'' = I_0 + 2I_1 + 2m_1\left(\frac{d}{8}\right)^2 + 2I_2 + 2m_2\left(\frac{3d}{8}\right)^2$$

Thus

$$I' - I'' = 2m_2\left\{\left(\frac{d}{8}\right)^2 - \left(\frac{3d}{8}\right)^2\right\} + 2m_1\left\{\left(\frac{3d}{8}\right)^2 - \left(\frac{d}{8}\right)^2\right\}$$

$$= (m_1 - m_2)\frac{d^2}{4}$$

Thus $I' - I''$ can be evaluated and τ determined from equation (5.63). If the radius of the wire, a, and its length l are accurately measured n can then be found, since

$$n = \frac{2\tau l}{\pi a^4} \qquad \qquad \dots (5.64)$$

5.18 DETERMINATION OF THE RIGIDITY MODULUS AND YOUNG'S MODULUS FOR THE MATERIAL OF A WIRE BY SEARLE'S METHOD
This method was developed by Searle in order to determine Young's modulus and the rigidity modulus for a material in the form of a short, fairly thin wire. The apparatus is shown in *Figure 5.13*. The wire specimen, *EF*, is firmly attached to two identical metal bars *AB* and *CD* at their mid points, the metal bars being suspended by parallel threads from their mid points *G* and *H*, such that the axis of suspension and the axis of the wire intersect at the centre of gravity of the bars. The bars are suspended at a distance apart such that in equilibrium the wire specimen is in a straight line. The ends *B* and *D* of the bars are drawn together and fastened by a loop of thin cotton, thereby causing the wire to be bent into an arc of a circle, radius *R*, subtending an angle of 2θ at the centre of curvature. Since the bars are suspended by threads the only couple acting on them is that due to the bending of the wire. The internal bending moment of the wire is $E\,Ak^2/R$. Now if the length of the wire is l, $R = l/2\theta$. Also if the wire is circular with radius a, $Ak^2 = \pi a^4/4$. The couple acting is therefore equal to

$$E.\frac{\pi a^4}{4}\cdot\frac{2\theta}{l} = \frac{\pi E a^4 \theta}{2l} \qquad \qquad \dots (6.65)$$

If the system is released by burning the cotton loop holding B and D, the relaxation of the strain in the wire causes each bar to oscillate in a horizontal plane. If the moment of inertia of a bar about its axis of suspension is I and its angular acceleration is $\ddot{\theta}$ the

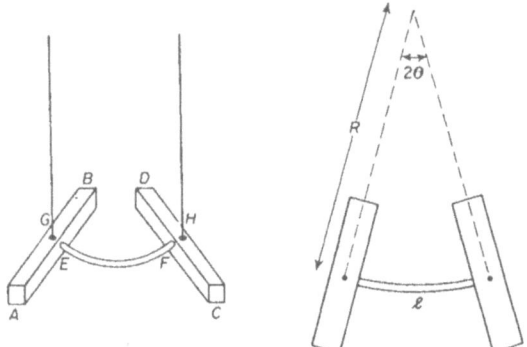

Figure 5.13. Determination of E and n by Searle's method

total external force is equal to $I\ddot{\theta}$. This is equal and opposite to the restoring couple exerted by the bent wire and hence the equation of motion for each bar is

$$I\ddot{\theta} = -\frac{\pi E a^4}{2l}\theta \qquad \ldots (5.66)$$

This is an equation of simple harmonic motion of which the period is given by

$$T_1 = 2\pi\sqrt{\frac{2Il}{\pi E a^4}} \qquad \ldots (5.67)$$

Hence by measurement of the period of the oscillations, E may be evaluated if I, l and a are known. For a bar of length $2b$, width $2a$ and mass m,

$$I = m \cdot \frac{a^2 + b^2}{3} \qquad \ldots (5.68)$$

The modulus of rigidity of the wire may be determined in a separate experiment as follows. The system is arranged with the wire specimen vertical, one of the bars, AB, being directly above the other, CD. The upper bar is firmly clamped and the lower bar displaced slightly from the equilibrium position. The wire is thereby twisted and thus causes a restoring couple to act on the bar which performs oscillations in a horizontal plane. If the angle of displacement is ϕ the restoring couple, due to the twisting of the wire, is $(\pi n a^4/2l)\,\phi$ as shown in Section 5.6. The equation of motion of the bar is therefore

$$I\ddot{\phi} = -\frac{\pi n a^4}{2l}\phi \qquad \qquad \dots (5.69)$$

Thus the bar oscillates with simple harmonic motion and the periodic time is given by

$$T_2 = 2\pi\sqrt{\frac{2Il}{\pi n a^4}}$$

Measurement of T_2 thus enables n to be evaluated.

It has previously been shown that $E = 2n(1 + \sigma)$, equation (5.10), and hence the two periodic times can be used to give a value for Poisson's ratio σ, since

$$2(1 + \sigma) = \frac{E}{n} = \frac{T_2^2}{T_1^2} \qquad \qquad \dots (5.70)$$

5.19 DETERMINATION OF YOUNG'S MODULUS FOR A METAL ROD BY AN INTERFERENCE METHOD

The apparatus is shown in *Figure 5.14*. The metal rod *AB* is placed vertically on a rigid support and a couple is applied at the end of

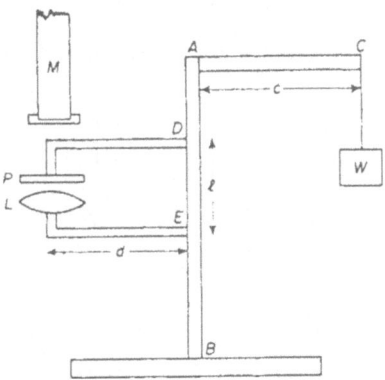

Figure 5.14. Determination of E *by interference method*

the rod by attaching a weight *W* to the horizontal arm *AC*. Due to the effect of this couple the metal rod is slightly curved and if the radius of curvature is *R* the internal bending moment in the rod is

133

EAk^2/R and this is equal to the external applied couple. The weight on the rod also causes it to be slightly compressed and if the rod has radius a, this compression is equal to $W/\pi a^2 E$ per unit length of rod. The effect of the curvature and compression of the rod is observed by attaching the rigid horizontal arms to the rod at D and E where $DE = l$. The arm at D carries a piece of plane glass P and the arm at E carries a lens L. By the usual optical arrangement Newton's rings are formed in the air film between the plane and curved glass surfaces and are observed through the microscope M.

Assume that the two glass surfaces are just touching when the rod is not acted upon by any load and let their separation be x when a load W is applied. Due to the curvature of the rod the upper arm is inclined to the lower arm at an angle θ, where $l = R\theta$, and the surfaces of the lens and plate become separated by an amount $d\theta$, where d is the distance of the centre of the Newton's ring system from the axis of the rod. The surfaces of the lens and plane glass are thus separated by the amount x where this is equal to the separation brought about by the curvature of the rod minus the distance they are brought together due to the compression of the rod, i.e.

$$x = d\theta - \frac{Wl}{\pi a^2 E}$$

$$\therefore \qquad x = \frac{dl}{R} - \frac{Wl}{\pi a^2 E} \qquad \qquad \dots (5.71)$$

but $EAk^2/R = Wc$ where c is the length of arm AC. Substituting in equation (5.71)

$$x = \frac{dlWc}{EAk^2} - \frac{Wl}{\pi a^2 E}$$

For a rod of circular cross-section $Ak^2 = \pi a^4/4$ and hence

$$x = \frac{Wl}{\pi a^4 E}(4cd - a^2) \qquad \qquad \dots (5.72)$$

In this experiment light from a sodium lamp is directed onto the plate and lens via an inclined plate. Newton's rings are brought into focus in the microscope and the diameter of the smallest dark ring, d_1, measured by means of a scale in the micrometer eyepiece. When the load is applied to the rod the rings shrink and disappear at the centre and the total number disappearing, n, are counted. When the apparatus has attained equilibrium the diameter of the smallest dark ring d_2 is again measured. If d_1 is not equal to d_2 the separation of the lens and plate has been increased by $(n + f) \lambda/2$ where n is an integer, f a fraction, and λ the wavelength of the light used. Hence

$$E = \frac{2Wl(4cd - a^2)}{\pi a^4 (n + f) \lambda} \qquad \dots (5.73)$$

Now the difference in the squares of the ring diameters is proportional to the phase difference of the rings. If the difference in the squares of the diameters S^2 of successive rings is measured, then

$$f = \frac{d_1^2 - d_2^2}{S^2}$$

and thus f may be evaluated and E determined from equation (5.73).

5.20 DETERMINATION OF YOUNG'S MODULUS AND POISSON'S RATIO FOR A GLASS BEAM

The bending of a beam was discussed in Section 5.7 when it was shown that the filaments of the beam on one side of the neutral surface are stretched while those on the other side are compressed. As a consequence of this extension and compression in the longitudinal direction, the lateral dimension of the beam is also changed, thereby leading to lateral curvature of the beam. With reference to *Figure 5.15(a)*, *AB* represents part of the neutral surface in a beam which is bent with a radius of curvature R_1. At a distance S below the neutral filament the filament will be compressed by an amount S/R_1 per unit length and hence there will be a lateral expansion at this level in the beam equal to $\sigma S/R_1$ where σ is Poisson's ratio. As a result of the change in lateral dimensions of the beam, its cross-section will have an appearance as shown in *Figure 5.15(b)*. In this direction the radius of curvature of the neutral surface of the beam is denoted by $O'C$. The expansion per unit length is $(C'D' - CD)/CD$ and the radius of curvature is $R'_1 = R_1/\sigma$.

If the beam is bent further so that the radius of curvature changes from R_1 to R_2 in the longitudinal direction and from R'_1 to R'_2 in the lateral direction, then Poisson's ratio is given by the equation

$$\sigma = \left(\frac{1}{R'_2} - \frac{1}{R'_1} 1 \right) \bigg/ \left(\frac{1}{R_2} - \frac{1}{R_1} \right) \qquad \dots (5.74)$$

The changes in the curvatures can be conveniently investigated by placing a lens on the beam as in *Figure 5.16* and observing Newton's rings which are formed in the air film between the lower surface of the lens and the beam *EFG*.

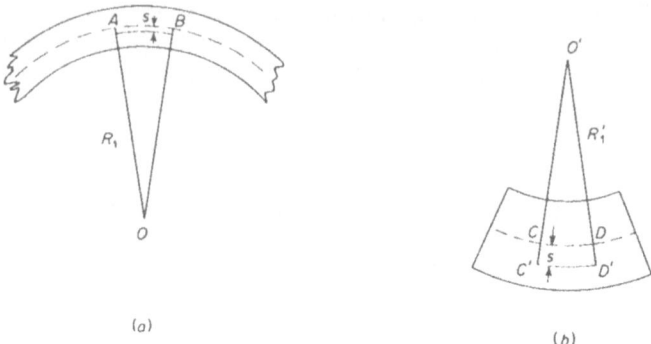

(a)

(b)

Figure 5.15. Determination of elastic constants for a glass beam

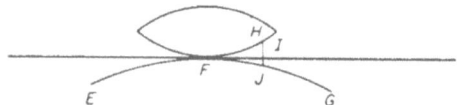

Figure 5.16. Formation of Newton's rings

At any point on the lens, such as H, some of the light rays are reflected and some are refracted to J where they are reflected, thus giving a path difference of $2HJ$ between the light reflected at H and at J. If the radius of curvature of the lower surface of the lens is R_0 and the longitudinal radius of curvature of the beam is R_1, then if the distance $FI = d$

$$HJ = HI + IJ = \frac{d^2}{2R_0} + \frac{d^2}{2R_1}$$

If there is to be reinforcement of the rays

$$\frac{d^2}{2}\left(\frac{1}{R_0} + \frac{1}{R_1}\right) = (n + c)\lambda \qquad \dots (5.75)$$

where λ is the wavelength of the light used and c is a constant to account for the phase change on reflection at J.

Similarly, if the transverse radius of curvature of the beam is R_1', then

$$\frac{d'^2}{2}\left(\frac{1}{R_0} - \frac{1}{R_1'}\right) = (n' + c)\lambda \qquad \dots (5.76)$$

136

The negative sign of $1/R_1$ is due to the transverse radius of curvature being in the opposite sense to that of the longitudinal radius of curvature.

The apparatus is set up as shown in *Figure 5.17* with the glass beam resting symmetrically on knife edges. Equal weights are

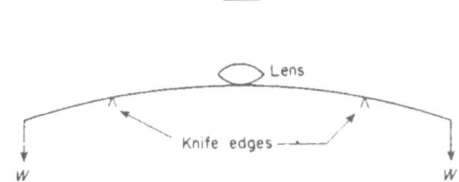

Figure 5.17. Determination of elastic constants for a glass beam

applied to the ends of the beam and by the usual optical arrangement Newton's rings are produced at the centre of the beam. The diameters of the rings in both the longitudinal and transverse directions are measured by means of a travelling microscope. Graphs are plotted of the squares of the radii of the rings, d^2 and d'^2, against their numbers n as abscissae. If the slopes of the graphs are θ_1 and θ'_1 respectively then, from equations (5.75) and (5.76)

$$\cot \theta_1 = \frac{1}{2\lambda}\left(\frac{1}{R_1} + \frac{1}{R_0}\right) \qquad \dots\,(5.77)$$

and
$$\cot \theta'_1 = \frac{1}{2\lambda}\left(\frac{1}{R_0} - \frac{1}{R'_1}\right) \qquad \dots\,(5.78)$$

If the weights on the ends of the beam are changed, then the radii of curvature change to R_2 and R'_2 respectively and the slopes of the graphs become θ_2 and θ'_2 respectively.

Now
$$\cot \theta_2 - \cot \theta_1 = \frac{1}{2\lambda}\left(\frac{1}{R_2} - \frac{1}{R_1}\right)$$

and
$$\cot \theta'_2 - \cot \theta'_1 = \frac{1}{2\lambda}\left(\frac{1}{R'_2} - \frac{1}{R'_1}\right) \qquad \dots\,(5.79)$$

and thus, from equation (5.74)

$$\sigma = \frac{\cot \theta'_2 - \cot \theta'_1}{\cot \theta_2 - \cot \theta_1} \qquad \dots\,(5.80)$$

137

If the external applied couple bending the beam is initially G_1 and finally G_2, then

$$G_2 - G_1 = EAk^2 \left(\frac{1}{R_2} - \frac{1}{R_1} \right)$$

$\therefore$ $$G_2 - G_1 = EAk^2 \, . \, 2\lambda(\cot \theta_2 - \cot \theta_1) \qquad \dots (5.81)$$

and thus E may be determined since for a beam of rectangular cross-section $Ak^2 = ab^3/12$, where a is the width and b the depth of the beam.

5.21 THE HELICAL SPRING

Figure 5.18 illustrates a section through a helical spring whose coils are inclined at an angle α to the horizontal. Let R be the radius of the cylinder on which the coils are wound. If the force causing the spring to be extended is W, then the external couple at any point A on the coils is WR. This couple can be resolved into a torque, $\Gamma = WR \cos \alpha$, acting in the plane of the section at A, and a bending moment, $G = WR \sin \alpha$, with its axis perpendicular to the section at A.

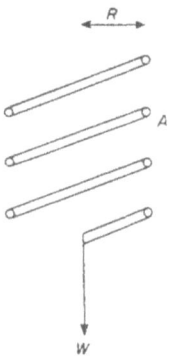

Figure 5.18. The helical spring

If the spring is extended through a distance s, the work done in stretching the spring is

$$V = \int_0^s W \, ds$$

Now this must be equal to the energy contained in the spring due to the bending, together with the energy contained in the spring due to the torsion, that is, from equations (5.20) and (5.28)

138

$$V = \frac{1}{2} \int \frac{G^2}{E A k^2} \, dl + \int \frac{\Gamma^2}{\pi n a^4} \, dl \qquad \ldots (5.82)$$

where $G = E A k^2 / R'$ for a beam bent with radius of curvature R'.

Hence

$$\int_0^s W \, ds = \frac{1}{2} \frac{G^2 l}{E A k^2} + \frac{\Gamma^2 l}{\pi n a^4} \qquad \ldots (5.83)$$

Substituting for G and Γ

$$\int_0^s W \, ds = \frac{1}{2} \frac{W^2 R^2 l \sin^2 \alpha}{E A k^2} + \frac{W^2 R^2 l \cos^2 \alpha}{\pi n a^4}$$

Differentiating with respect to s

$$\frac{ds}{dW} = R^2 l \left(\frac{\sin^2 \alpha}{E A k^2} + \frac{2 \cos^2 \alpha}{\pi n a^4} \right) \qquad \ldots (5.84)$$

Since $W = 0$ when $s = 0$, then

$$s = W R^2 l \left(\frac{\sin^2 \alpha}{E A k^2} + \frac{2 \cos^2 \alpha}{\pi n a^4} \right) \qquad \ldots (5.85)$$

For a wire of circular cross-section, $A k^2 = \pi a^4 / 4$ and hence

$$s = \frac{2 W R^2 l}{\pi a^4} \left(\frac{2 \sin^2 \alpha}{E} + \frac{\cos^2 \alpha}{n} \right) \qquad \ldots (5.86)$$

For a flat helical spring, i.e. $\alpha = 0$,

$$s = \frac{2 W R^2 l}{\pi n a^4} \qquad \ldots (5.87)$$

Besides the vertical extension of the free end of the spring there is an angular displacement in the horizontal plane. If the end of the wire is twisted through ϕ the torsion gives rise to a horizontal angular displacement, β, equal to $\phi \sin \alpha$. Since

$$\Gamma = W R \cos \alpha = \frac{\pi n a^4}{2l} \phi$$

$$\beta = \frac{2 W R l \sin \alpha \cos \alpha}{\pi n a^4} \qquad \ldots (5.88)$$

This causes the spring to coil up since it acts inwards.

139

The bending moment, on the other hand, produces a horizontal angular rotation of the free end, γ, where

$$\gamma = \int_0^l \frac{dl \cos \alpha}{R^1} = \frac{WR \sin \alpha \cos \alpha \cdot l}{EAk^2} = \frac{4WRl \sin \alpha \cos \alpha}{E \cdot \pi a^4}$$

$$\dots (5.89)$$

(since $R' = G/EAk^2$ and $G = WR \sin \alpha$). This causes the spring to uncoil, since it acts outwards, and thus the total angular displacement as the spring coils up is given by $\beta - \gamma$, i.e.

$$\frac{2WRl \sin \alpha \cos \alpha}{\pi a^4} \left(\frac{1}{n} - \frac{2}{E} \right) \qquad \dots (5.90)$$

This has a maximum value when $\alpha = 45°$ and it can be seen that the spring will coil up or uncoil as $1/n \gtrless 2/E$. For most metals $E > 2n$ and helical springs generally coil up when stretched.

5.22 VERTICAL OSCILLATIONS OF A LOADED SPRING

For a flat spring, i.e. $\alpha = 0$, only the torsional energy need be considered. When the spring is subjected to a couple of magnitude WR then the potential energy V is given by

$$V = \frac{W^2 R^2 l}{\pi n a^4} \qquad \dots (5.91)$$

Combining equations (5.87) and (5.91)

$$V = \frac{\pi n a^4}{4l R^2} x^2 \qquad \dots (5.92)$$

If the velocity of the moving mass is dx/dt when the extension of the spring is x, then kinetic energy of the mass is $\frac{1}{2} \cdot W/g(dx/dt)^2$. If the free end of the wire of the spring moves with the velocity dx/dt then the kinetic energy of an element ds of the wire at a distance s from the fixed end is $\frac{1}{2}m(s/l \cdot dx/dt)^2 ds$ where m is the mass per unit length of the spring wire and l is the total length of the spring wire. The total kinetic energy of the spring is therefore obtained by integration

$$\int_0^l \frac{1}{2}m \left(\frac{dx}{dt} \right)^2 \frac{s^2}{l^2} ds = \frac{1}{6} \frac{w}{g} \left(\frac{dx}{dt} \right)^2 \qquad \dots (5.93)$$

where w is the weight of the spring. The total kinetic energy of the spring and load is then

$$\tfrac{1}{2}g\left(W + \frac{w}{3}\right)\left(\frac{dx}{dt}\right)^2 \qquad \ldots (5.94)$$

Applying the law of conservation of energy

$$\tfrac{1}{2}g\left(W + \frac{w}{3}\right)\left(\frac{dx}{dt}\right)^2 + \frac{\pi n a^4}{4lR^2}\cdot x^2 = \text{constant} \quad \ldots (5.95)$$

Differentiating with respect to t

$$\left(W + \frac{w}{3}\right)\frac{d^2x}{dt^2} + \frac{\pi n a^4 g}{2lR^2}\cdot x = 0 \qquad \ldots (5.96)$$

This is an equation of simple harmonic motion and the period of vibration is given by

$$T = 2\pi \sqrt{\frac{W + w/3}{\pi n a^4 g/2lR^2}} \qquad \ldots (5.97)$$

5.23 STRAIN GAUGES

The electrical resistance strain gauge is based on the fact that if a metal wire is extended or compressed its resistance changes, the change in resistance being proportional to the change in length. It is illustrated in *Figure 5.19* and consists of a long length of resistance

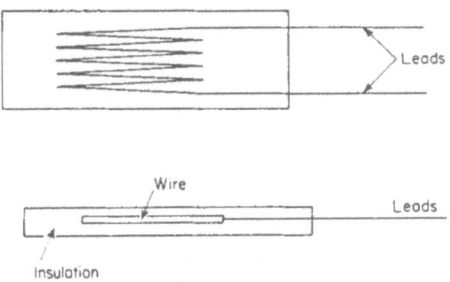

Figure 5.19 Strain gauge

wire arranged to form a grid. The grid so formed is placed within an insulating medium and the strain gauge is then firmly cemented to the surface to be tested. The theory of the gauge is as follows. If

141

F*

the resistivity of the strain gauge wire is ρ, the resistance of the wire of length l and radius a is given by

$$R = \frac{\rho l}{\pi a^2} \qquad\qquad \ldots\text{(5.98)}$$

$\therefore$ $\qquad\qquad \log R = \log \rho - \log \pi + \log l - 2\log a$

hence $\qquad\qquad \dfrac{dR}{R} = \dfrac{dl}{l} - \dfrac{2da}{a} \qquad\qquad \ldots\text{(5.99)}$

Now by definition Poisson's ratio σ is $-(da/a)/(dl/l)$, and thus

$$\frac{dR}{R} = \frac{dl}{l}(1 + 2\sigma)$$

$(1 + 2\sigma)$ is referred to as the strain sensitivity of the gauge and hence the strain dl/l can be written as

$$\frac{dl}{l} = \frac{dR}{R} \cdot \frac{1}{\text{strain sensitivity}} \qquad\qquad \ldots\text{(5.100)}$$

To measure the small changes in resistance, the strain gauge is connected in one arm of a Wheatstone bridge as shown in *Figure*

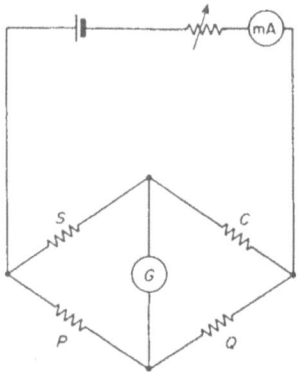

Figure 5.20. Strain gauge bridge

5.20. In this figure, S represents the strain gauge and P and Q are standard resistances which can be varied. C is a compensating gauge and is constructed to be identical to the actual strain gauge but is left free from strain. It is subjected to the same temperature changes as S, and thus neutralizes any change in resistance of the gauge with temperature.

The strain gauge is extremely useful for measuring the strains in flat or curved objects due to its relatively small size, usually a few centimetres long. Once the equipment has been calibrated it also provides a rapid direct measurement of strain.

5.24 THE COMPRESSIBILITY OF SOLIDS AND LIQUIDS

The compressibility of solids and liquids is extremely small and it was not until 1762 that Canton first demonstrated that water was compressible. The first experiments to yield accurate information on the compressibilities of liquids were not carried out until 1847 when Regnault performed a series of experiments; his equipment is described later. A few years later, in 1877, Amagat used an improved form of piezometer to determine the compressibility of various liquids and worked at up to 3000 atm. The name piezometer has been given to the apparatus used for carrying out these measurements at high pressure. Experiments at a higher pressure were limited by the difficulty of overcoming leaks in the apparatus at such high pressures and it was not until Bridgman designed a new technique to overcome these leaks that experiments at higher pressures were possible. At present, experiments at pressures of the order of 20,000 atm are possible and some important results have been obtained on the behaviour of matter at such pressures.

5.25. THE COMPRESSIBILITY OF LIQUIDS

The early work of Regnault and Amagat was carried out with piezometers. In their experiments the containing vessel was subject to a change in volume as well as the liquid under test and their results for the compressibility of liquids are subject to an uncertainty in the change in volume of the container. The type of apparatus used is shown diagrammatically in *Figure 5.21*. The liquid under test is contained in the bulb and capillary tube, the upper end of which is connected to a compressor pump and a manometer. Pressure can be applied to the outside of the bulb through the liquid in the outer container, the compressor pump being connected to this liquid via the tap A. The taps A, B and C enable the pressure to be applied to the inside of the bulb only, to the outside of the bulb only, or to both inside and outside simultaneously.

If the apparent change in volume of the liquid under test is δV_a when a pressure P is applied simultaneously to the inside and outside of the bulb, the true change in volume will be

$$\delta V = \delta V_a + \delta V_c \qquad \dots (5.101)$$

143

where δV_c is the decrease in the internal volume of the bulb. In the case of a cylinder of isotropic material with flat ends it can be shown that

$$\frac{\delta V_c}{V_c} = \frac{P}{K_c} \qquad \qquad \ldots (5.102)$$

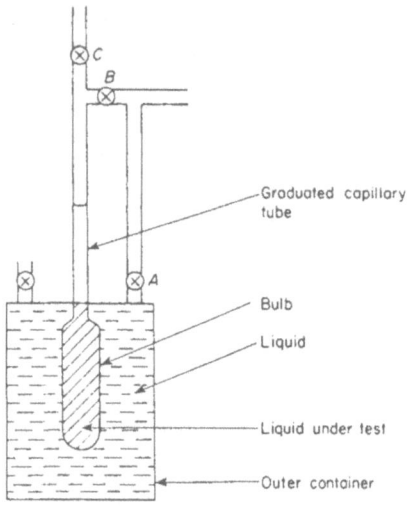

Figure 5.21. Piezometer

where K_c is the bulk modulus of the cylinder material. Thus, provided K_c is known, the change in the bulb volume δV_c can be calculated. If the bulk modulus of the liquid under test is K then it can be calculated according to the equation

$$\frac{\delta V_a}{V_c} = P\left(\frac{1}{K} - \frac{1}{K_c}\right) \qquad \qquad \ldots (5.103)$$

Regnault actually used as a container a cylinder with rounded ends and his results are therefore subject to some inaccuracy.

Modern work on the compressibility of liquids has been done by Bridgman who has carried out measurements at pressures of up to 20,000 atm. *Figure 5.22* illustrates the type of apparatus used by Bridgman which was designed to withstand very high pressures. The liquid under examination is contained in a hardened steel case.

Pressure is applied via the steel piston. This is not directly in contact with the liquid, but is connected *via* a steel ring pressing on the soft rubber packing enclosed between copper washers which in turn presses on the mushroom shaped steel head. It can thus be seen that, due to this method of transmitting the pressure to the liquid, no

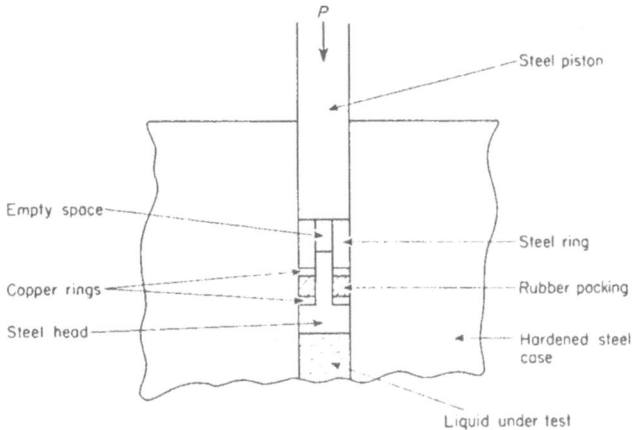

Figure 5.22. Bridgman's apparatus

liquid can leak past the packing since the pressure along the sides is always greater than the upward pressure from the liquid. The pressure on the rubber packing is greater than that in the liquid in the ratio of the area of the steel head to the area of the rubber ring.

With this type of apparatus Bridgman obtained the standard pressure–volume isothermals over a wide range of temperature. The pressure was measured by a manganin resistance and this is described in more detail in a later section. Bridgman found that as the pressure increased the compressibility of liquids decreases considerably and, for instance, at a pressure of 12,000 atm the compressibility is only about 1/20 of its value at moderate pressures. This can be explained by assuming that at moderate pressures the molecules fit loosely together with an amount of free space available. The compressibility will be quite large until this has been used up but after this any further volume changes could be due to the shrinkage of the molecules themselves.

5.26 THE BULK MODULUS OF SOLIDS

Several techniques were used by early workers in an attempt to measure the bulk modulus of solids directly. One method used by

Amagat is shown in *Figure 5.23*. It consists of a vertical cylindrical tube of the material under examination which has attached to it varying loads. The change in the volume of the cylinder is measured by the movement of a liquid contained in the cylinder. This movement is observed by means of the open capillary tube fixed to the top of the cylinder. Theoretically, in the case of a cylindrical tube, it can be shown that the change in volume of the cylinder is given by

$$\frac{\delta V_c}{V_c} = \frac{P(1 - 2\sigma)}{E} = \frac{P}{3K_c} \qquad \ldots (5.104)$$

where P is the applied stress, σ is Poisson's ratio and $\delta V_c / V_c$ is the volume strain.

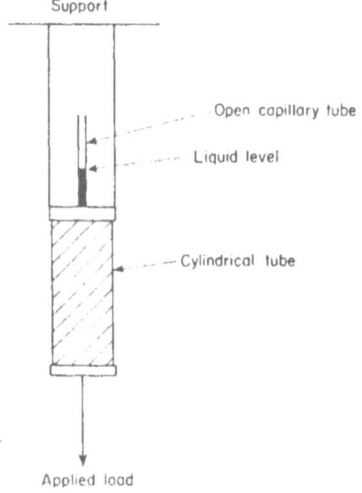

Figure 5.23. *Amagat's apparatus*

For modern accurate work the technique developed by Bridgman is the most reliable. The apparatus he designed is shown schematically in *Figure 5.24*. The material under examination is in the form of a solid cylindrical rod and is contained in a heavy steel cylinder. This cylinder is then immersed in a high pressure chamber as shown in *Figure 5.22* and is subjected to an external pressure applied hydrostatically. The specimen contracts and its contraction relative to the steel cylinder, l_1, is determined by the movement of a loose-fitting ring which moves to a new position during the contraction and subsequently remains in this new position even after the pressure is removed. During the experiment the cylinder extends in length slightly, this change in length being fairly small compared with the contraction of the specimen. However, if this change in length is l_2 then the true contraction of the specimen is $l = l_1 - l_2$. The change in length of the cylinder is determined by comparator measurements and thus the true change in length of the specimen can be evaluated. The volume strain is equal to three times the longitudinal strain. By this method it is possible to obtain the absolute value of the compressibility of the specimen.

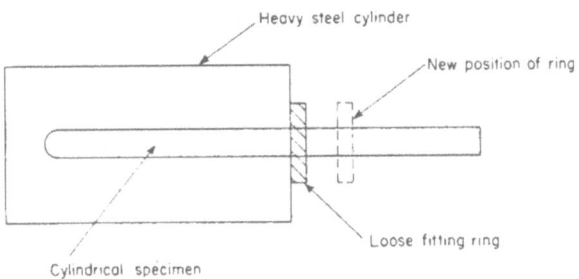

Figure 5.24. Bridgman's apparatus for measuring the absolute compressibility of solids

Apparatus for enabling the measurement of relative compressibilities to be made was also designed by Bridgman and is shown in *Figure 5.25*. In this equipment the material under test is in the form

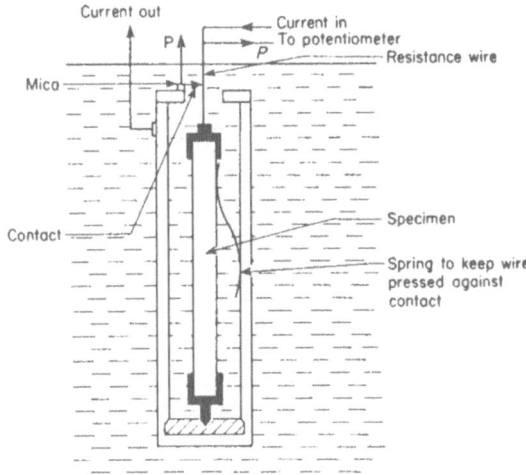

Figure 5.25. Bridgman's apparatus for determining relative linear compressibilities

147

of a long rod and is pressed firmly against the bottom of the iron container by a spring. A high resistance wire sliding over a contact insulated from the container, is attached to the upper end of the rod. The relative position of the container and the wire is found by a measurement of the potential difference between the sliding contact and a terminal fixed to the wire. The whole apparatus is subjected to hydrostatic pressure and hence the relative linear compressibility is directly determined from the change in resistance, the current being kept constant.

5.27 COMPRESSIBILITY OF GASES

Real gases obey closely the normal gas law

$$PV = RT \qquad \dots (5.105)$$

provided the pressure is not excessively great and the temperature excessively low. From this equation it can be shown that, under isothermal conditions, the bulk modulus of the gas is given by

$$K_T = - V\left(\frac{dP}{dV}\right)_T = P \qquad \dots (5.106)$$

Under adiabatic conditions, where $PV^\gamma = $ constant, the adiabatic elasticity is given by

$$K_H = - V\left(\frac{dP}{dV}\right)_H = \gamma P \qquad \dots (5.107)$$

A large number of experimenters investigated the relationship between the pressure and volume of real gases and the work of Regnault and Amagat is particularly important. Their experiments are fully described in various textbooks and are not discussed further here. It was established that for pressures exceeding a few thousand atm, there is no basic difference in compressibility between gases and liquids and, for instance, air is as dense as water at such pressures. The large compressibility initially observed in the case of gases is due to the relatively large distances between the molecules. However, at pressures of a few thousand atmospheres where the density of gases approaches that of liquids, any further decrease in volume is possibly due to a decrease in the volume of the molecules themselves, and the compressibility observed is consequently much smaller than the initial value.

As a result of the experimental results of Regnault and Amagat many attempts have been made in order to obtain a general expression relating the pressure, volume and temperature of a substance over a wide range of pressure. These attempts have led to such well-known equations as van der Waals', Clausius', Dieterici's.

Bridgman has carried out experiments at pressures of up to 16,000 atm. and his apparatus is shown in *Figure 5.26*. In this apparatus the pressure can be steadily increased on the gas contained in the cylinder. The piston is steadily moved down, increasing the pressure in the kerosene until it equals the pressure of the gas. The

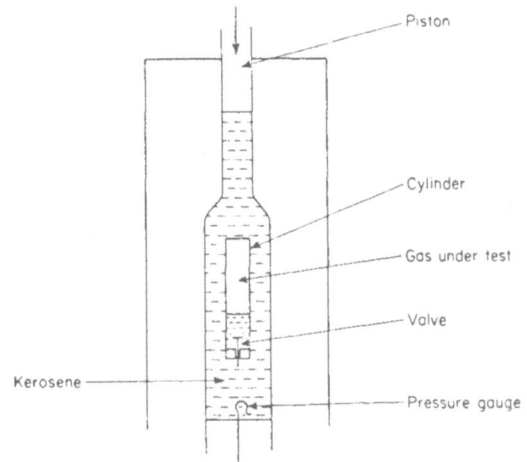

Figure 5.26. Bridgman's apparatus for examining gases at high pressures

valve is then forced open and the piston movement observed. The change in volume, due to the fact that both the kerosene and gas are compressible, can then be deduced. Prior to this stage the piston motion had been determined only by the apparent compressibility of the kerosene. Thus the compressibility of the gas can be calculated. Bridgman found that for nitrogen, hydrogen and helium the product PV is almost a linear function of the pressure.

5.28 THE MEASUREMENT OF HIGH PRESSURES

The simplest form of primary pressure gauge is the mercury manometer. Obviously such a gauge has a fairly low upper limit set by the height of the column and in practice a mercury manometer is only used for pressures of up to a few hundred atmospheres.

For pressures exceeding 1000 atm the only type of primary gauge which can be used is the free piston gauge which was first used by Amagat. Basically it consists of a piston of small cross-sectional area exposed directly on one side to the pressure which is to be measured. The piston slides in a cylinder accurately fitted, so that there are virtually no leaks along the sides. The pressure is measured directly from the load which must be applied to the piston in order to maintain it in equilibrium. To eliminate the effect of friction the piston is rotated before each measurement is made.

For work at very high pressures it is not usually convenient to use the free piston gauge and consequently it is standard practice to use a secondary gauge. Secondary gauges can utilize any convenient effect which is brought about by high pressures and a particularly useful effect is the variation of electrical resistance of a metallic wire with change of pressure. Bridgman showed that soft metal wires could be used, and developed the manganin wire gauge to such an extent that it is now the standard method of measuring very high pressures.

Such a gauge consists of a length of annealed manganin wire having a normal resistance of about 120 Ω. This gauge is calibrated against a primary gauge up to a pressure of about 7000 atm and Bridgman established that up to this pressure the change of resistance of manganin shows a linear variation with the change in pressure. For work at higher pressures the calibration graph can be extrapolated, although this may result in some error.

6

SURFACE TENSION

6.1 INTRODUCTION

It is well known that, in many ways, a liquid behaves as though its surface were covered with an elastic skin. For instance, it is possible to make a needle 'float' on the surface of water, although the needle does not float in the usual sense of the term and it sinks immediately if the skin of the surrounding water is broken. Small drops of a liquid also tend to become spherical in shape even though they are acted upon by gravity and would thus be expected to become a flat surface. All liquids behave in this manner and it is usual to explain such behaviour on the basis of the molecular forces existing between the constituent molecules of the liquid.

Molecules in the interior of the liquid, well removed from the liquid surface, are subjected to the attractive force of the other liquid molecules in all directions so that the resultant of these forces is very small on average and the molecule does not rapidly move from its position. However, molecules at the surface of the liquid are subjected to assymetric fields of force because of the net force of attraction by molecules in the bulk of the liquid. Hence there is a net force tending to draw molecules at the surface into the interior of the liquid, where there is no appreciable resultant force. As a result of the asymmetric field of force exerted on surface molecules the surface of a liquid contracts until it contains the minimum number of molecules possible. Thus for any given volume of liquid the surface area is the smallest possible, taking into account the effect of external forces. All the molecules at the surface of the liquid, therefore, possess potential energy by virtue of their position and the amount of this potential energy per unit area of surface is termed the surface energy density, E.

A simple illustration of surface energy is provided by the behaviour of a soap film. Consider a loop of thread mounted on a wire frame, the frame being covered with a soap film (*Figure 6.1*). If the soap film is broken inside the loop, the remaining film pulls the thread out to form a circle. A circle has the largest area for a given perimeter and hence it is clear that the soap film contracts as much as possible.

This experiment shows that the contraction of a liquid surface is associated with a decrease of potential energy.

While the surface energy of a liquid is of fundamental importance, the term surface tension, T, is more frequently encountered. This term has been in common use for a considerably longer period than

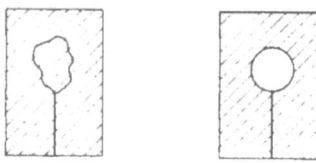

Figure 6.1. Behaviour of a soap film

the term surface energy. Surface tension was associated with the idea of the liquid surface being in a state of tension with forces acting parallel to the surface. Such forces cannot be explained on the basis of the molecular forces which lead to the idea of forces acting perpendicularly to the liquid surface. In dealing theoretically with the formation and shapes of liquid surfaces, however, it is often convenient to utilize this hypothetical surface tension and it is defined as the force per centimetre exerted in the tangent plane to the liquid surface in a direction normal to an element of a line drawn in the surface through any point. It is assumed to have the same value at every point in a given liquid surface whatever the shape of the surface.

It is important to realize that the definitions of surface tension and surface energy are not equivalent. This is because an allowance must be made for the numerical difference between a force tending to produce a deformation and the energy change associated with the deformation. Surface tension actually decreases with increasing temperature. Hence, if a liquid surface were extended adiabatically the temperature would drop. If it is extended isothermally then it absorbs heat from its surroundings. The change in energy of the surface, therefore, is not exactly equal to the work done.

The relationship between surface energy density and surface tension may be derived on a thermodynamic basis as follows. Suppose a soap film is mounted on a frame with a moveable boundary of length l. First the film is stretched a distance x at a temperature θ_1. The work done in moving the boundary through the distance x is $T_1 lx$, where T_1 is the value of the surface tension at the temperature θ_1. The increase in area of the film is lx and thus the energy change is Elx. Hence the heat absorbed by the film from its surroundings is

$(Elx - T_1 lx)$. The film is then cooled to a temperature θ_2, the area remaining constant. In this process no work is done but an amount of heat is given out, $C(\theta_1 - \theta_2)$, where C is the heat capacity of the film. Following this, the film is allowed to contract at the temperature θ_2 until its area has decreased by an amount lx. The work done by the film is $T_2 lx$ where T_2 is the surface tension at the temperature θ_2. Finally the film is heated to the temperature θ_1, its area remaining constant. Again no work is done but an amount of heat $C'(\theta_1 - \theta_2)$ is absorbed.

The net result of the complete thermodynamic cycle is that an amount of heat, $Elx - T_1 lx$, has been absorbed at the temperature θ_1 and an amount of mechanical work equal to $(T_2 - T_1) lx$ has been done by the film. The above cycle is reversible and hence, neglecting the difference between C and C', according to the laws of thermodynamics

$$\text{Efficiency} = \frac{\text{Work done}}{\text{Heat absorbed at temperature } \theta_1} = \frac{\theta_1 - \theta_2}{\theta_1}$$

Hence,

if $\dfrac{\theta_1 - \theta_2}{\theta_2}$ is small

$$\frac{(T_2 - T_1) lx}{(E - T_1) lx} = \frac{\theta_1 - \theta_2}{\theta_1}$$

In the limit as $\theta_1 - \theta_2 \to 0$

$$E - T = - \theta \frac{\partial T}{\partial \theta} \qquad \dots (6.1)$$

From equation (6.1) it is apparent that the surface tension is only equal to the surface energy density at the absolute zero when $\theta = 0$, or if $\partial T / \partial \theta = 0$. Now the surface tension decreases with increase in temperature for all liquids so $\partial T / \partial \theta$ is negative, hence E is always greater than T.

6.2 PRESSURE ON A CURVED FILM OF CONSTANT SURFACE TENSION

Let $ABCD$ be a small portion of the curved film having radii of curvature r_1 and r_2. Let the excess pressure on the concave side counterbalancing the surface tension effect be equal to p Nm^{-2}. Suppose the surface area xy is displaced normally through a distance

dz so that it occupies the position $A'B'C'D'$, the temperature remaining constant. The work done by the pressure is then equal to $p.xy.dz$. The increase in surface area of the film is equal to $\delta(xy)$ and the work done is given by $2T.\delta(xy)$, since there are two surfaces to the membrane.

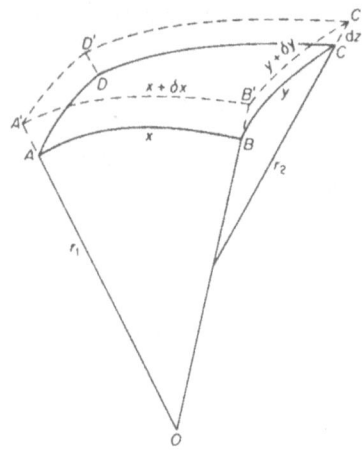

Figure 6.2. Pressure on a curved film of constant surface tension

In *Figure 6.2* it can be seen that triangles ABO and $A'B'O$ are similar, so that

$$\frac{x + \delta x}{r_1 + \delta z} = \frac{x}{r_1} = \frac{\delta x}{\delta z}$$

Thus
$$\delta x = \frac{x}{r_1}\delta z$$

Similarly
$$\delta y = \frac{y}{r_2}\delta z$$

Now
$$p.xy.dz = 2T\delta(xy)$$
$$= 2T(x\delta y + y\delta x)$$

hence
$$p.xy.dz = 2T\left(\frac{xy\delta z}{r_2} + \frac{xy\delta z}{r_1}\right)$$

therefore
$$p = 2T\left(\frac{1}{r_1} + \frac{1}{r_2}\right) \qquad \dots\,(6.2)$$

154

In the case of a spherical soap bubble where $r_1 = r_2 = r$, the excess pressure inside the bubble is given by $p = 4T/r$ according to equation (6.2).

If only one liquid surface is considered, e.g. a liquid drop or an air bubble in a liquid, then the corresponding expression is

$$p = T\left(\frac{1}{r_1} + \frac{1}{r_2}\right) \qquad \dots (6.3)$$

6.3 CONTACT OF SOLIDS, LIQUIDS AND GASES

Consider a system consisting of two non-miscible liquids A and B. Suppose the liquids are initially in contact and that they are then separated completely by a direct pull. Let the mechanical work per square metre required to separate them be W_{AB} joules. Before separation there is potential energy at the interface between the two liquids, the amount per unit area being known as the interfacial energy density. After separation there is potential energy at the surfaces of the two liquids.

Let the surface energy densities of the interface and of the liquids be E_{AB}, E_A and E_B, respectively, and let the corresponding surface tensions be T_{AB}, T_{AG} and T_{BG}. Let the amounts of heat supplied, per unit area of the new surfaces formed, in order to maintain iso-thermal conditions be H_A and H_B, and let H_{AB} be the corresponding quantity for the interface. Since the interface has been destroyed and two liquid–air surfaces have been created in the process, the total heat supplied per unit area of surface is $-H_{AB} + H_A + H_B$. After separation of the liquids, the total surface energy density is $E_A + E_B$. Applying the principle of conservation of energy

$$E_{AB} + W_{AB} - H_{AB} + H_A + H_B = E_A + E_B \qquad \dots (6.4)$$

But it was seen earlier that $E = T + H$ and hence, substituting in equation (6.4)

$$T_{AB} + H_{AB} + W_{AB} - H_{AB} + H_A + H_B = T_{AG} + H_A + T_{BG} + H_B \qquad \dots (6.5)$$

and hence
$$W_{AB} = T_{AG} + T_{BG} - T_{AB} \qquad \dots (6.6)$$

Equation (6.6) is known as Dupré's equation.

Consider next a system consisting of a solid, a liquid and a gas, all in contact and in equilibrium as shown in *Figure 6.3*. Suppose them to meet along a common line of contact so that the molecules in this line are in equilibrium. The forces acting on the molecules in this line are T_{LG}, T_{SL} and T_{SG} per unit length, according to the definition of surface tension. Let ϕ be the angle between the tangent planes to the liquid and to the solid along their line of contact. This is known as the contact angle for the particular liquid and solid.

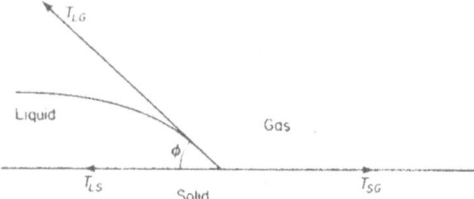

Figure 6.3. Liquid–solid–gas system

Since the net horizontal force per unit length to the left is equal to the net horizontal force per unit length to the right, then

$$T_{SL} + T_{LG} \cos \phi = T_{SG} \qquad \dots (6.7)$$

But, from equation (6.6), for the liquid-solid system

$$W_{SL} = T_{LG} + T_{SG} - T_{SL} \qquad \dots (6.8)$$

Eliminating the unknown solid-liquid and solid-gas surface tensions, T_{SL} and T_{SG} from equations (6.7) and (6.8), gives

$$W_{SL} = T_{LG}(1 + \cos \phi) \qquad \dots (6.9)$$

Equation (6.9) is known as Young's equation. W_{SL} is a measure of the attraction between the solid and the liquid, i.e. the adhesion. If Dupré's equation is written for a liquid in contact with itself, where there is obviously no interfacial surface tension, then

$$W_{LL} = 2T_{LG} \qquad \dots (6.10)$$

Hence $2T_{LG}$ is a measure of the adhesion of the liquid itself, i.e. its cohesion. Now if the angle of contact between a solid and a liquid is zero, then from equation (6.9)

$$W_{SL} = 2T_{LG} \qquad \dots (6.11)$$

A comparison of equations (6.10) and (6.11) shows that if ϕ is 0, the adhesion between the liquid and the solid is equal to the cohesion of the liquid.

6.4 NEUMANN'S TRIANGLE

Consider a drop of liquid A to be placed on the surface of liquid B, the two liquids being non-miscible. Suppose the drop of liquid does not spread over the other liquid but remains as a drop. Let the point of contact between the two liquids and air be O as shown in *Figure 6.4*. The line through O is subject to the three forces T_{AG}, T_{BG} and T_{AB} per unit length, corresponding to the surface tensions of the free surfaces of liquids A and B and the surface tension of the interface respectively. If such a situation were feasible then it would be possible to draw a triangle of forces with the three sides proportional to the three surface tensions. This triangle is known as Neumann's triangle.

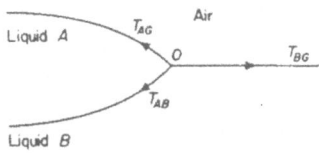

Figure 6.4. Neumann's triangle

In practice it has never yet been found possible to draw such a triangle. This means that one liquid always spreads over the surface of the other. However, many examples can be quoted where a drop of one liquid will rest, apparently in equilibrium, on another liquid surface. For instance a drop of benzene appears to remain on the surface of water without spreading. Such behaviour is normally explained by assuming that the surface of one of the liquids is made impure by contamination with the other and that when a drop of benzene is apparently resting on water, it is not really resting on the water surface but on a thin film of benzene which has spread over the water. Other examples, such as water on a mercury surface, can also be accounted for by the contamination of one of the surfaces. In fact, in this latter example, if the mercury is absolutely clean, water spreads over it quite easily.

6.5 ANGLE OF CONTACT BETWEEN LIQUIDS AND SOLIDS

The form of a liquid surface in the vicinity of a solid is dependent on the forces exerted on each liquid molecule by the liquid and solid molecules in its vicinity. *Figure 6.5* illustrates three types of behaviour which may be observed. In case (*a*) the molecules of the solid attract the liquid molecules very strongly indeed so that the angle of contact is very small and the liquid is said to 'wet' the solid, e.g. pure water on clean glass. In case (*b*) the liquid tries to 'wet' the solid but does not completely succeed. This is known as 'imperfect wetting', and there is a finite angle of contact, e.g. water on a glass surface contaminated with grease. In case (*c*) the angle of contact is obtuse, and in such a case the liquid is said not to 'wet' the solid, e.g. mercury on glass.

In general, there is an angle of contact for every liquid at rest on a solid, and this parameter, together with the surface tension of the liquid, is necessary in order to describe the behaviour of a liquid in the neighbourhood of different solids. Sometimes it is found that

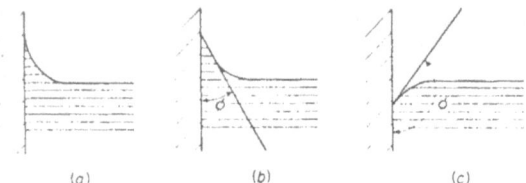

Figure 6.5. Angle of contact between liquids and solids

the angle of contact is not constant but varies depending on whether the liquid is advancing or receding over the surface of the solid. In such a case the angles of contact are specified as the advancing and receding angles of contact, ϕ_A and ϕ_R respectively. Such behaviour may be exhibited by mercury on glass.

6.5.1 Determination of angle of contact by method of Adam and Jessop

The angle of contact between a solid and a liquid can be determined directly by the method of Adam and Jessop if the solid is available in the form of a flat plate. The apparatus is shown in *Figure 6.6*. It consists of a rectangular plate glass trough which has the tops of the sides ground flat and coated with a suitable non-contaminating substance, e.g. paraffin wax, if the liquid being investigated is water. The flat plate is held in a clamp so as to dip into the liquid and it can be rotated about a horizontal axis. The trough is filled to the top with the liquid.

Before an experiment is carried out the liquid surface is swept free of contamination by moving a scraper coated with non-contaminating material across its surface. The plate is then lowered into the liquid at various angles of inclination and when the angle of setting is correct the liquid continues horizontal right up to the plate. After each setting of the plate a period of about 1 min must elapse to allow the surface to regain its equilibrium. If ϕ is the angle between the edge of the plate and the horizontal surface of the water then this is the angle of contact for the solid and liquid concerned. Adam and Jessop considered it sufficiently accurate to measure the angle with a protractor, since the angle varies slightly from point to point on the plate.

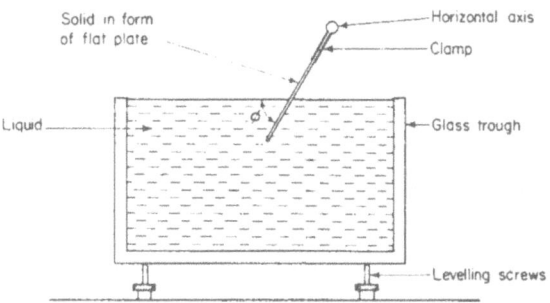

Figure 6.6. Angle of contact by method of Adam and Jessop

6.5.2 Determination of angle of contact by method of Ablett

Another method for measuring the angle of contact between paraffin wax and water is due to Ablett. The measurements take into account the fact that the angle of contact varies according to whether the liquid is advancing or receding over the solid or is stationary. The effect of varying the speed of the liquid relative to the solid can also be investigated.

The apparatus is shown in *Figure 6.7(a)*. It consists of a solid glass cylinder about 8 cm in diameter and 8 cm long covered with a smooth layer of paraffin wax. The cylinder is mounted so that it can rotate about its axis, its axis being horizontal and parallel to the longer sides of a rectangular glass tank containing water. A suitable gearing system is provided to enable the linear surface velocity of the cylinder to be varied up to a speed of about 4 mm/s. The cylinder is partly immersed in the water and facilities are provided for adding or removing water from the tank, thereby varying the depth of immersion of the cylinder. One end of the tank is covered with dull black paper in which a fine horizontal slit S_2 is made. A beam of parallel light from a suitable source is allowed to pass obliquely upwards through an adjustable slit S_1, through the slit S_2, and on to the underside of the water surface, from where it is reflected downwards towards the observer. Adjustment of S_1 and the optical system for producing the parallel beam enables the beam of light to be directed to any desired point on the water surface near to the curved surface of the cylinder.

The image generally seen by the observer is shown in *Figure 6.7(b)*, the horizontal image of the slit normally being curved near the

159

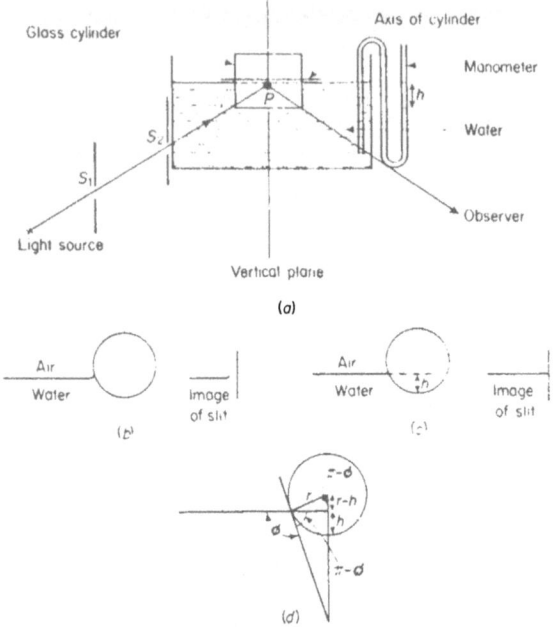

Figure 6.7. Angle of contact by Ablett's method

cylinder. *Figure 6.7(b)* also shows the general end view of the cylinder partly immersed in the water. The depth of water in the tank is varied, thus varying the depth of immersion of the cylinder, the optical arrangement and the slit S_1 being adjusted so that the point P is always in the vertical plane. The water depth is adjusted until the point is reached when the water surface remains horizontal right up to the surface of the cylinder, as shown in *Figure 6.7(c)*. Under these conditions the image of the slit also appears completely horizontal, as shown. The depth of immersion, h, of the cylinder in this position, is noted. An accurate method of measuring h is by using a manometer in conjunction with the apparatus, the difference in manometer readings being noted when the slit image is completely horizontal and when the water is just touching the cylinder.

In *Figure 6.7(d)*, if the radius of the glass cylinder is r and the angle of contact is ϕ, then

$$-\cos \phi = \cos (\pi - \phi) = \frac{r - h}{r}$$

$$\cos \phi = \frac{h - r}{r} \qquad \ldots (6.12)$$

160

Thus, measurement of h and knowledge of r enables ϕ to be calculated.

Experiments are carried out with the cylinder stationary and also when it is slowly rotating in clockwise and anti-clockwise directions. In each case the water depth in the tank is adjusted until the slit image is completely horizontal. Anti-clockwise rotation observed from the left-hand side of the cylinder corresponds to the water advancing over the paraffin wax, similarly, clockwise rotation corresponds to the water receding. Ablett's results established that within experimental error

$$\phi = \frac{\phi_A + \phi_R}{2} \qquad \text{.... (6.13)}$$

ϕ_A and ϕ_R being the 'advancing' and 'receding' angles of contact respectively.

He also showed that for cylinder speeds up to 0·44 mm/s, the angles of contact varied in a definite way but that at all speeds equation (6.13) was satisfied. At speeds greater than 0·44 mm/s the angles of contact were constant.

6.6 FORCE BETWEEN TWO PARALLEL PLATES SEPARATED BY A THIN LAYER OF LIQUID

It is well known that if two parallel glass plates are separated by a thin layer of liquid, then although they slide over each other quite easily, a considerable force must be exerted to separate them normally. The situation is illustrated in *Figure 6.8*, where the two plates of radius R are a distance d apart. Let the surface tension of the liquid be T and the angle of contact between the liquid and the glass be ϕ.

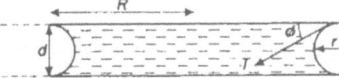

Figure 6.8. Force between two parallel plates separated by a thin layer of liquid

Now, according to equation (6.3), the pressure inside the liquid is less than that outside by an amount $T[(1/r) - (1/R)]$ where r is the radius of curvature of the meniscus as shown and is equal to $d/2\cos\phi$. The normal component of the force, due to surface tension, acting round the periphery of a plate is $2\pi RT \sin\phi$. Thus the resultant force acting downwards on the upper plate is

$$T\left(\frac{1}{r} - \frac{1}{R}\right)\pi R^2 + 2\pi RT \sin\phi$$

Hence, substituting for r, the force required to separate the plates is given by

$$F = 2\pi R T \left(\frac{R \cos \phi}{d} - \frac{1}{2} + \sin \phi \right) \qquad \dots (6.14)$$

For most liquids which wet plates, the latter two terms in the bracket may be neglected and the equation becomes

$$F = \frac{2\pi R^2 T \cos \phi}{d} \qquad \dots (6.15)$$

It is evident that for plane surfaces where d is very small, F can be very large indeed. In fact, if an attempt is made to pull the plates apart normally, there is considerable risk that the plates may break.

6.7 FORCE REQUIRED TO PULL A PLATE NORMALLY FROM A LIQUID SURFACE

Normally, when a flat plate is placed on the surface of a liquid and slightly raised, a layer of liquid will be pulled upwards with the plate, as shown in *Figure 6.9*. If the force raising the plate is gradually increased, the amount of liquid being raised also increases until the plate breaks away. Let the maximum height to which the liquid is raised before the plate breaks away, be h and let the force exerted at this point be F. Let atmospheric pressure be P_0 and let the density of the liquid be ρ. Then the pressure in the liquid just below the plate, at Q, is $(P_0 - \rho g h)$.

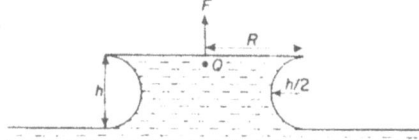

Figure 6.9. Force required to pull a plate normally from a liquid surface

If the plate is assumed to be circular, of radius R, the resultant force pulling the plate downwards is $\pi R^2 [P_0 - (P_0 - \rho g h)] = \pi R^2 \rho g h$ and this is equal to F. If R is large it can be assumed that the shape of the meniscus in a vertical plane is a semi-circle of radius $h/2$. At a point where the horizontal cross-section of the elevated liquid is a minimum, the pressure difference across its surface is approximately

$$T \left(\frac{1}{r} - \frac{1}{R} \right) \simeq \frac{T}{r} \simeq \frac{2T}{h}$$

162

From the previous assumption that in the vertical plane the meniscus has a semi-circular shape, this pressure difference is equal to $\rho gh/2$. Hence, $h^2 = 4T/\rho g$, so that

$$F = \pi R^2 \rho gh = 2\pi R^2 \sqrt{\rho g T} \qquad \dots (6.16)$$

If this force is measured by means of a balance then a value for T can be obtained. It must be remembered, however, that equation (6.16) is only approximately true.

6.8 CAPILLARY ASCENT

When a clean capillary tube is placed vertically with its lower end below the surface of a liquid, it is well known that the liquid rises in the tube to a level above that of the level of the liquid outside. The height to which the liquid rises depends on factors such as its surface tension, angle of contact, etc., and can be evaluated as follows. Let the liquid, of density ρ and surface tension T, have risen to a height h in the capillary tube of radius r (*Figure 6.10*). Let the angle of contact between the liquid and material of the tube be ϕ and let R be the radius of curvature of the liquid surface at all points.

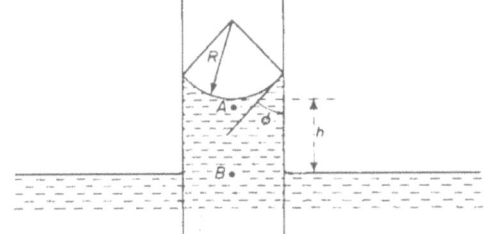

Figure 6.10. Capillary ascent

The pressure over the curved surface of the liquid is atmospheric, P_0. At A, a point just below the liquid surface, the pressure is less than atmospheric by an amount $2T/R$, according to equation (6.3). At B, a point in the same horizontal plane as the liquid surface outside the tube, the pressure is also atmospheric. The difference in pressure between the two points A and B is equal to the pressure exerted by a column of liquid of height h, ρgh. Hence the pressure at B equals pressure at A plus ρgh, i.e.

$$P_0 = P_0 - 2T/R + \rho gh$$

so that $\qquad 2T/R = \rho gh$

But $r = R \cos \phi$, and hence

$$2T \cos \phi = r\rho gh \qquad \dots (6.17)$$

In the case of liquids such as water, alcohol, etc., the angle of contact is assumed to be zero so that the formula is simplified to

$$T = \frac{r\rho gh}{2} \qquad \dots (6.18)$$

In the formulae derived above, the assumption was made that the liquid surface was hemispherical with constant radius of curvature R. This assumption is only justified when the capillary tube is very narrow and the formulae must be modified accordingly in all other cases. Suppose the form of the meniscus in the plane of the diagram is an ellipse with semi-axes r and b (*Figure 6.11*). The radius of curvature at the point A is then $R = r^2/b$. Now $2T/R = \rho gh$, and if $T/\rho g$ is put equal to a^2, then

$$2a^2 = Rh \qquad \dots (6.19)$$

Substituting for R

$$2a^2 = r^2h/b \qquad \dots (6.20)$$

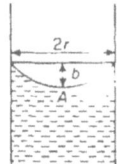

Figure 6.11. Ascent in a wide capillary tube

If the tube is assumed to be perfectly cylindrical, then the force due to surface tension round the perimeter of the liquid surface is equal to the total weight of the liquid column. Hence, assuming the angle of contact to be zero

$$2\pi r T = \pi r^2 h\rho g + \tfrac{1}{3}\pi r^2 b\rho g$$

Substituting for b from equation (6.20) gives

$$2a^2 = rh \left(1 + \frac{r^2}{6a^2}\right)$$

or $$12a^4 - 6a^2rh - r^3h = 0 \qquad \dots (6.21)$$

164

Solving equations (6.21) gives

$$a^2 = \frac{6rh \pm 6rh\left(1 + \frac{4}{3}\cdot\frac{r}{h}\right)^{\frac{1}{2}}}{24}$$

Since, in the limit, when $r/h \to 0$, $2a^2 = rh$, the positive sign must be taken and hence

$$\frac{T}{\rho g} = \frac{rh}{2}\left(1 + \frac{1}{3}\frac{r}{h} - \frac{1}{9}\frac{r^2}{h^2} + \ldots\right) \qquad \ldots (6.22)$$

This formula gives results which are very accurate for values of r up to $0.2\,h$.

6.9 RISE OF A LIQUID BETWEEN VERTICAL PLATES

If vertical plates are parallel at a small distance d apart, as in *Figure 6.12*, the liquid surface can be assumed to be a section of a cylindrical surface. The pressure at A, just below the liquid surface, is less than atmospheric by an amount $T(1/r + 1/\infty)$, according to equation (6.3), but $d = 2r$ and hence $T/r = 2T/d$. Assuming the angle of contact to be zero, then as shown previously in Section 6.8, this pressure difference is equal to $\rho g h$, where ρ is the liquid density and h the height of the liquid between the plates above the liquid level outside. Hence

$$\frac{2T}{d} = \rho g h \qquad \ldots (6.23)$$

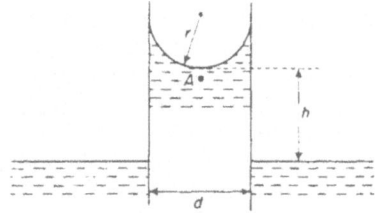

Figure 6.12. Liquid ascent between vertical parallel plates

If the vertical plates are inclined to each other at an angle θ the liquid rises between the plates as shown in *Figure 6.13*. The curves of contact of the liquid, of density ρ, with the plates are NHL and NGK. Considering the equilibrium of the element of liquid bounded by the vertical planes $FGHI$ and $JKLM$, the

165

downward force due to its weight is equal and opposite to the resultant upward force due to surface tension. The weight of the element is approximately equal to $FI . FJ . GF . \rho g$ dynes. If $OF = x$, $FG = y$ and $FJ = dx$, and if θ is assumed to be small so that $FI = x\theta$, then the weight of the element is $xy\theta dx . \rho g$.

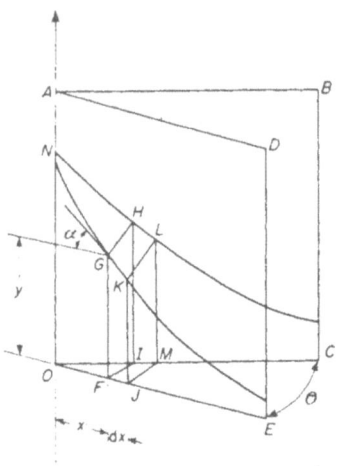

Figure 6.13. Liquid ascent between vertical inclined plates

The forces due to surface tension act on the edges GH, HL, LK and GK. To a first approximation the surface tension forces due to the adjacent liquid acting along GH and KL can be assumed to be equal and opposite and may be neglected. Let $GK = ds$ and let the tangent to the curve at G make an angle α with the horizontal. The surface tension forces acting on GK and HL act along the normals to GK and HL and each has the value $T ds$. The resultant vertical force is thus $2T ds \cos \alpha$, but $ds \cos \alpha = dx$, and hence the resultant vertical force upwards is $2T dx$. Hence, to a first approximation,

$$xy\theta dx\rho g = 2T dx$$

i.e. $$xy = 2T/\rho g\theta \qquad \qquad \dots (6.24)$$

The curve NGK is thus approximately a rectangular hyperbole. In the foregoing theory the angle of contact of the liquid with the

166

plates has been assumed to be zero and the upward force due to the air displaced by the liquid has been neglected.

6.10 METHODS OF MEASUREMENT OF SURFACE TENSION
6.10.1 Capillary ascent method

The apparatus is shown in *Figure 6.14* and consists of a uniform clean capillary tube clamped in a vertical position. There are various methods of cleaning the capillary but to obtain a perfectly clean bore capillary it is probably best to draw it out in a flame immediately prior to the experiment. In order to reduce the error of the experiment it is advisable to use several freshly made capillaries and take the average value of all the measurements. The vertical tube is placed with one end below the surface of the liquid being investigated and is thoroughly wetted, by immersing in the liquid to a greater depth than that to which the liquid will rise during the experiment. It is then clamped in position with one end below the liquid surface, the liquid level outside the tube being indicated by a suitable marking device attached to the capillary.

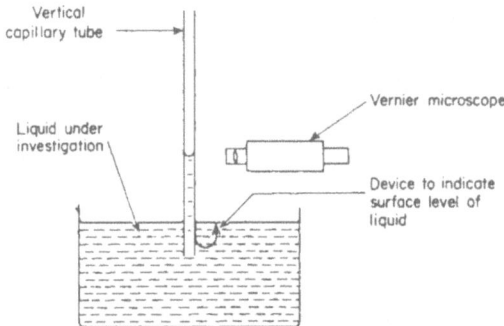

Figure 6.14. Measurement of surface tension by capillary rise method

The height to which the liquid rises in the capillary is measured by a vernier microscope. After measuring this height the capillary is then broken at the point where the top of the liquid meniscus was and the radius of the tube is found by measuring the diameter with the vernier microscope in two directions at right angles. The value of the surface tension of the liquid may then be found from the formulae derived in Section 6.8, assuming the density of the liquid to be known.

While the capillary rise method is widely used for the measurement of surface tension, the result is not usually very accurate for a number of reasons. Probably the most important reason for inaccuracy is the difficulty of ensuring that the tube is absolutely clean and has a uniform cross-section. Since surface tension varies with temperature, the temperature of the liquid meniscus should also be accurately measured, but this is rarely done. It is also common practice to assume the angle of contact to be zero, but when it is not, then $T \cos \phi$ should be used for greater accuracy.

6.10.2 Jaeger's method

Basically, Jaeger's method consists of measuring the maximum pressure required to produce an air bubble at the end of a vertical capillary tube which is immersed in the liquid whose surface tension is to be measured. It overcomes most of the inherent errors of the capillary rise method. The apparatus is shown in *Figure 6.15*. The capillary tube, of radius r, is placed vertically in the liquid of density ρ_1, under investigation, so that the end of the capillary is at a distance x below the surface of the liquid. The capillary is connected to a manometer containing xylol or some similar liquid of density ρ_2, and also to a vessel of large volume, which acts as a reservoir and is connected to a pressure pump. The tap is adjusted until bubbles are formed at the end of the capillary at the rate of one every few seconds. The maximum pressure difference of the liquid levels in the manometer is noted, and hence the maximum pressure in the bubbles before they break away from the capillary can be found.

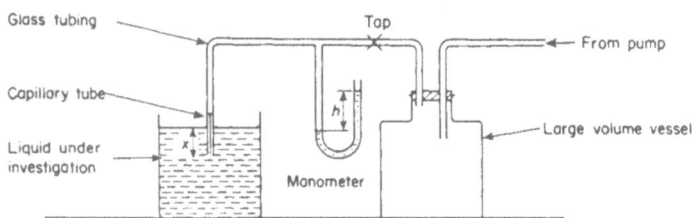

Figure 6.15. Measurement of surface tension by Jaeger's method

The maximum pressure p_1 inside the bubbles is given by

$$p_1 = P_0 + \rho_2 g h$$

where P_0 is the atmospheric pressure. Now the pressure in the liquid just outside the bubble is

$$p_2 = P_0 + \rho_1 g x$$

168

If it is assumed that the maximum pressure inside the bubble is reached when the bubble reaches a hemispherical shape of radius r equal to the radius of the capillary, then the excess pressure inside the bubble over that outside. at the point of breaking away, is $2T/r$.

Hence
$$\frac{2T}{r} = p_1 - p_2 = \rho_2 gh - \rho_1 gx$$

and
$$T = \frac{gr}{2}(\rho_2 h - \rho_1 x) \qquad \ldots (6.25)$$

If r, ρ_2, ρ_1, and x are known and h is measured, then T can be calculated.

Jaeger's apparatus provides a simple means of investigating the variation of the surface tension with the temperature of the liquid. In such an investigation, for accurate results, the variation of the liquid density with temperature must also be considered. This information may not always be available but a simple variation of Jaeger's experiment, due to Sugden, overcomes this difficulty. In Sugden's modification, two capillary tubes, of different radii, are used consecutively in Jaeger's experiment and the maximum pressures recorded by the manometer are measured in each case. If the capillaries are adjusted so that their ends are at the same distance below the liquid surface then, from equation (6.25)

$$\frac{2T}{r_1} = g(\rho_2 h_1 - \rho_1 x) \qquad \text{and} \qquad \frac{2T}{r_2} = g(\rho_2 h_2 - \rho_1 x)$$

Combining these equations gives

$$2T\left(\frac{1}{r_1} - \frac{1}{r_2}\right) = g\rho_2(h_1 - h_2) \qquad \ldots (6.26)$$

The size of the bubbles at maximum pressure may be governed by either the internal or external radius of the tube and hence it is more correct to write equation (6.25) as

$$T = f(r)\frac{g}{2}(\rho_2 h - \rho_1 x) \qquad \ldots (6.27)$$

where the equilibrium position of the bubble is some definite function $f(r)$ of the capillary radius. Since $f(r)$ is the same for all experiments with the same tube, equations (6.27) can be used in an investigation of the variation of surface tension with temperature.

6.10.3 Balance method

The apparatus is shown in *Figure 6.16* and consists of an ordinary balance, to one arm of which is attached a perfectly clean glass slide so that its lower edge is horizontal. A microscope slide is usually used for this purpose. The liquid being examined is contained in a suitable vessel and it completely fills the vessel so that the liquid surface can be swept perfectly clean by a scraper, as described in Section 6.5. With the glass plate poised over the liquid surface but not actually touching it, masses are added to the other pan of the balance until equilibrium is attained. The vessel containing the liquid is then gently raised by means of the screw table until the

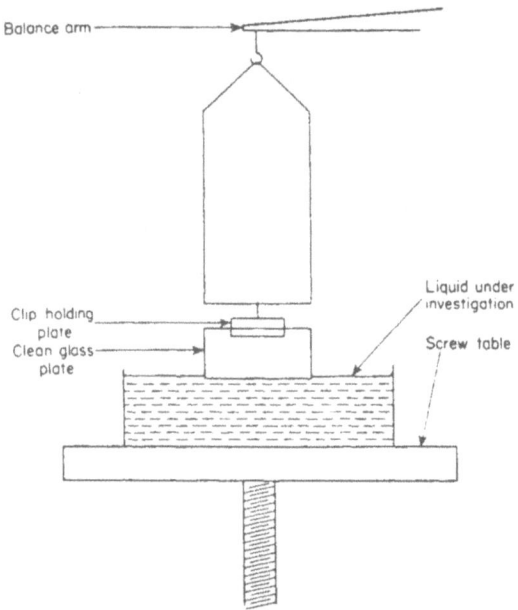

Figure 6.16. Measurement of surface tension by balance method

liquid surface just touches the lower edge of the plate. The plate is immediately pulled down and masses are added to the other pan of the balance until the plate is just withdrawn from the liquid. If the additional mass added to the balance pan to restore equilibrium,

170

is *m* and if the length and thickness of the glass plate are *l* and *t* respectively, then, since the surface tension force acts round the perimeter of the plate touching the liquid surface

$$2T(l + t) = mg \qquad \ldots (6.28)$$

The ordinary balance is not very suitable for surface tension measurements as, since the masses are added in finite amounts, however small the amount, there is a small jerk which may cause the meniscus to break prematurely. An alternative method is provided by the torsion balance which provides a means of increasing the load continuously without jerks. The torsion balance is illustrated in *Figure 6.17* and consists of a beam *AB*, rigidly attached at *O* to a torsion wire which is stretched between two firm supports, *C* and *D*. The large mass at *A* is used to adjust the beam to a horizontal position after which its position is securely fixed by a screw. The other end of the beam *B*, moves over a scale and if masses are suspended from the beam, it can always be adjusted to be in a horizontal position by twisting the torsion wire.

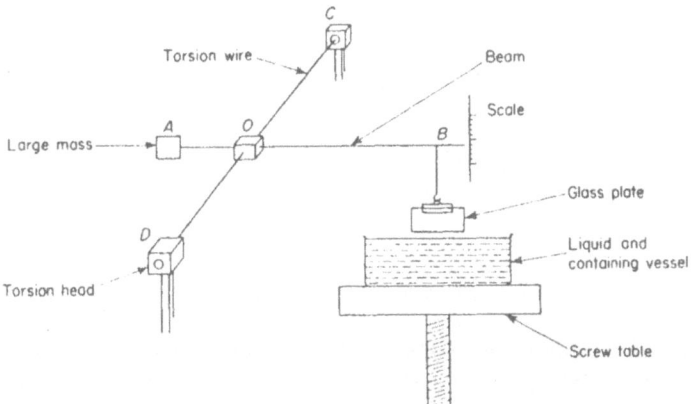

Figure 6.17. Measurement of surface tension by torsion balance method

In an experiment to measure surface tension, the liquid is placed in a vessel so as to completely fill it and the surface is cleaned by a scraper as described above. The microscope slide is suspended from the end of the beam and the torsion head is adjusted until the beam is horizontal. The liquid surface is then raised until the plate just touches the liquid surface. The torsion head is then adjusted until the plate just leaves the liquid surface and the difference in the readings of the torsion head is noted. Provided the torsion balance has been calibrated by suspending masses from the end of the beam the force due to surface tension can be evaluated and T found from equation (6.28).

6.10.4 Ferguson's method

Ferguson's method has the advantages of requiring only a small quantity of liquid and also a knowledge of liquid density is not necessary. It can also be used for the measurement of the interfacial tension of two liquids. The apparatus is shown in *Figure 6.18(a)* and consists of a horizontal clean capillary tube connected to a sensitive manometer, and a device to enable the pressure in the apparatus to be gradually increased.

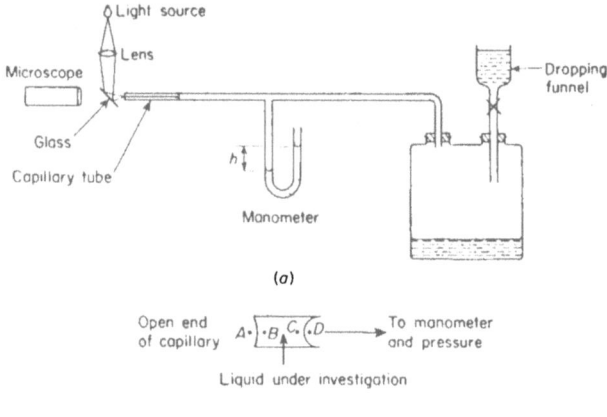

Figure 6.18. Measurement of surface tension by Ferguson's method

A small quantity of the liquid being investigated is placed in the capillary tube and, in the absence of a pressure difference between its ends, the two meniscus surfaces are similar. If a small excess pressure is applied to one end of the liquid surfaces, by allowing water to flow from the dropping funnel into the reservoir, then the liquid will be moved to the end of the capillary tube as shown in *Figure 6.18b*. A continued increase in the pressure causes the meniscus to change from concave, through plane, to a convex form. This technique affords an extremely sensitive method of determining the exact moment when the meniscus becomes plane. The pressure in the apparatus at any instant is given by the difference in heights, h, of the manometer liquid of density ρ. The meniscus at the open

172

end of the capillary is used as a reflecting surface for the small light source shown. If the radius of curvature of the meniscus is small an image of the light source is formed near to the pole of the meniscus but as the liquid surface becomes plane the image broadens and at the instant of planeness, a practically uniform field is observed through the microscope.

Suppose the excess pressure in the apparatus at any instant is $p = \rho g h$. At B and C inside the liquid the pressures are equal. However, the pressure at B is greater than that at A by $-2T/R$, where T is the surface tension of the liquid and R is the radius of curvature of the liquid surface at B. Similarly, the pressure at D is greater than that at C by $2T/r$, where r is the radius of curvature of the liquid surface at C. Hence the pressure at D is greater than that at A by $2T(1/r - 1/R)$ and this is equal to $\rho g h$. To measure the surface tension of the liquid in the capillary the pressure in the apparatus is adjusted until the liquid surface at the open end of the capillary becomes plane. Under these conditions the radius of curvature at B is infinity and hence $1/R = 0$. If the difference in heights of the liquid surfaces in the manometer at this point is H, then

$$\frac{2T}{r} = \rho g H \qquad \qquad \ldots . (6.29)$$

Thus, provided r, which is the radius of the capillary, is known, T may be determined. In the foregoing it has been assumed that the angle of contact between the liquid and the capillary is very small and also that the diameter of the capillary is sufficiently small so that the effect of gravity on the liquid meniscus is negligible.

6.10.5 Sentis' method

The apparatus consists of a perfectly clean capillary tube drawn out to a fine jet. The tube is dipped into the liquid being investigated and some of the liquid is drawn up into the tube. The tube is drawn from the liquid and the liquid in the tube falls gradually to form a small spherical drop at the end of the capillary as shown in *Figure 6.19(a)*. Now the pressure at A is greater than that at B by $2T/R$, where R is the radius of curvature of the liquid surface at B. Similarly, the pressure at C is greater than that at D by $2T/R_1$, where R_1 is the radius of curvature of the liquid surface at C. Hence, if the distance between B and C is h_1, then

$$\frac{2T}{R} + \frac{2T}{R_1} = \rho g h_1 \qquad \qquad \ldots . (6.30)$$

where ρ is the density of the liquid.

173

G*

If the drop is assumed to be spherical in shape below its maximum cross-section, then the radius of curvature at C, R_1, is equal to r, where $2r$ is the diameter of the drop at its maximum cross-section. Thus, measurement of $2r$ by a suitable travelling microscope enables R_1 to be determined. When this measurement is completed,

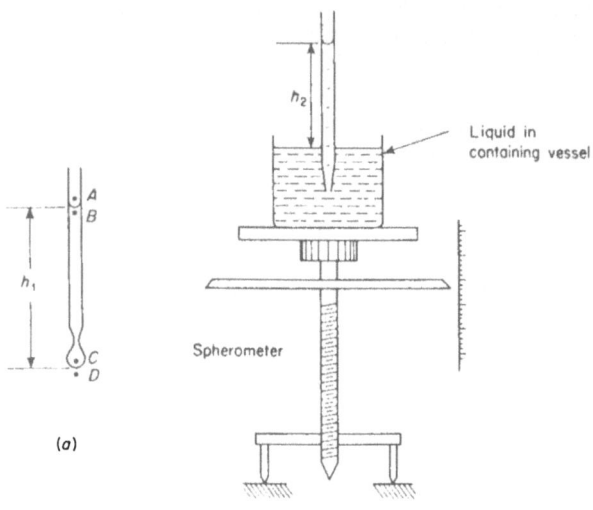

Figure 6.19. Measurement of surface tension by Sentis' method

a vessel containing the liquid is placed on the table of a spherometer underneath the drop. The table is raised until the liquid surface in the vessel just touches the drop, at which point the position of the table is observed. The vessel is then raised further until the upper surface of the liquid in the capillary is once again at its original position, as shown in Figure 6.19(b). The position of the table is again noted at this point when

$$\frac{2T}{R} = \rho g h_2 \qquad \qquad \dots (6.31)$$

where h_2 is the height of liquid in the capillary over the liquid surface in the vessel. Combining equations (6.30) and (6.31)

$$\frac{2T}{R_1} = \rho g(h_1 - h_2) \qquad \qquad \dots (6.32)$$

174

and, since $h_1 - h_2$ is the distance through which the table has been raised, T can be determined.

6.10.6 Drop-weight method

This method of measuring surface tension is very simple since it consists of measuring the mass of a liquid drop falling slowly from the tip of a vertical tube. The theory of the method assumes that as the drop breaks away it has a cylindrical form where it is attached to the tube. This situation is illustrated in *Figure 6.20* where the liquid, of surface tension T, is shown falling from a tube of radius r.

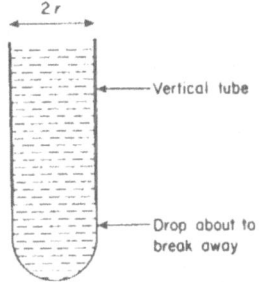

Figure 6.20. Measurement of surface tension by drop-weight method

Considering the static equilibrium of the drop just before it breaks away, the downward forces are due to its weight mg, and the excess pressure inside the drop acting over the area πr^2, i.e. $T/r \cdot \pi r^2$, since the excess pressure inside a drop of cylindrical cross-section is T/r. These forces are balanced by the surface tension forces acting upwards, therefore

$$mg + \frac{T}{r} \cdot \pi r^2 = 2\pi r \cdot T \qquad \qquad \text{....(6.33)}$$

Hence
$$mg = \pi r T \qquad \qquad \text{....(6.34)}$$

Unfortunately, the above assumption is not valid and the breaking away of the drop is really a dynamical problem.

If it is assumed that the weight of the drop is some function of r, T, ρ and V, where ρ and V are the liquid density and the volume of a drop respectively, dimensional analysis shows that

$$\frac{mg}{Tr} = \phi \left[\frac{r}{V^{\frac{1}{3}}} \right] = f \left[\frac{r}{B} \right] \qquad \qquad \text{....(6.35)}$$

where $B = \sqrt{(T/\rho g)}$, ϕ is some function of $r/V^{\frac{1}{3}}$ and f is some function of r/B. Using liquids of known surface tension and density, B is determined and values of mg/Tr, r/B and $r/V^{\frac{1}{3}}$ are obtained. Tubes of different radii are also used. Curves are drawn connecting mg/Tr with $r/V^{\frac{1}{3}}$ and with r/B, and these curves may subsequently be used to determine the surface tension of any liquid. In a measurement, m and r are determined, so that $r/V^{\frac{1}{3}}$ can be calculated, mg/Tr is determined from the graph, and hence T is calculated.

For accurate results several precautions must be taken. The tip of the tube used must be very flat and have sharp edges and this is usually achieved by precision grinding techniques. The formation of the drop must take place very slowly over several minutes, although this time can be reduced by artificially forming a large part of the drop fairly rapidly and then allowing its final growth and fall to take place under gravity alone. An experiment usually consists of collecting about 30 drops over a suitable period of time. Temperature control is achieved by placing the apparatus in a thermostatically controlled atmosphere.

This technique can be applied to the measurement of interfacial surface tension between two liquids and, in such an experiment, the weight of a drop of one liquid, of density ρ_1, falling from a tube which dips below the surface of a second liquid, of density ρ_2, is found. Under these circumstances equation (6.34) becomes

$$\frac{mg(\rho_1 - \rho_2)}{\rho_1} = \pi r T \qquad \ldots (6.36)$$

The dynamical problem is solved in an exactly similar manner to the case outlined above and similar curves may be derived.

6.10.7 Determination of surface tension by measurements on large sessile drops and bubbles

Figure 6.21(a) shows a large sessile drop of liquid resting on a horizontal plate, the medium above the drop being a gas, while *Figure 6.21(b)* shows a bubble of gas resting under a horizontal plate in a liquid medium. The drop is assumed to be large so that it is horizontal at its highest point and its diameter is large compared with its thickness, so that at any point on its surface the curvature, in a plane other than that in the plane of the diagram, is negligible.

Consider a point P on the surface of the drop, with coordinates (x, y), and let the tangent to the section at P make an angle ψ with the x axis. If the atmospheric pressure is P_0 then the pressure at P is equal to $P_0 + \rho g y$, where ρ is the density of the liquid. Also, if the radii of curvature at P are R_1 and R_2, in, and perpendicular to, the plane of the diagram, then the pressure at P is also equal to $P_0 + T(1/R_1 + 1/R_2)$. Since $1/R_2$ is negligible then, approximately

$$T/R_1 = \rho gy \qquad \ldots (6.37)$$

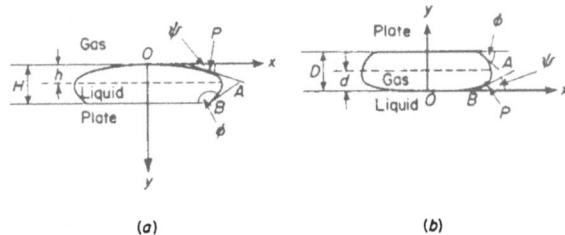

(a)　　　　　　　　　　(b)

Figure 6.21. (a) Sessile drop; (b) gas bubble

The curvature, $1/R_1$, is equal to $d\psi/ds$, where ds is a small element of the periphery of the drop at P in the plane of the diagram. Hence

$$\rho gy = T\frac{d\psi}{ds} = T \cdot \frac{d\psi}{dy} \cdot \frac{dy}{ds} = T\frac{d\psi}{dy}\sin\psi$$

Therefore $\rho g \int_0^y y\,dy = T\int_0^\psi \sin\psi\,d\psi$

and

$$\frac{\rho gy^2}{2} = T[1 - \cos\psi] \qquad \ldots (6.38)$$

If A, at a depth h below the x axis, is the point where the tangent is vertical, then since $\psi = 90°$

$$\rho gh^2/2 = T \qquad \ldots (6.39)$$

At the point of contact, B, ψ is equal to the angle of contact ϕ, and y is equal to the thickness of the drop H. Hence

$$\rho gH^2/2 = T[1 - \cos\phi] \qquad \ldots (6.40)$$

Thus, from equations (6.39) and (6.40), measurement of h and H enables both T and ϕ to be determined.

In the case of the gas bubble, it is convenient to have the y axis reversed as shown. As above, at any point P on the surface

$$\rho gy = T\frac{d\psi}{dy}\sin\psi$$

Therefore $\rho g \int_0^y y\,dy = T\int_0^\psi \sin\psi\,d\psi$

and hence, if A, height d above the x axis is the point where the tangent is vertical and if the thickness of the drop is D

$$\rho g d^2/2 = T \qquad \qquad \dots (6.41)$$

and

$$\rho g D^2/2 = T[1 + \cos \phi] \qquad \qquad \dots (6.42)$$

where ϕ is the angle of contact.

The formulae derived above are not sufficiently exact for accurate work since the assumption that $1/R_2$ is negligible produces errors which depend on the size of the drop. It should also be noted that the formulae were derived assuming that the angle of contact for the sessile drop was always such that $\pi/2 < \phi < \pi$. This is true for liquids which do not wet the surface, such as mercury on glass or water on paraffin wax. Similarly, in the case of the gas bubble, it was assumed that $0 < \phi < \pi/2$. If these conditions are not fulfilled, then h and d cannot be found. H and D can be measured, but T can only be found if ϕ is known, or vice versa.

In an experiment to determine the surface tension of mercury, a large drop of clean mercury is formed on a clean horizontal glass plate and H and h are measured by a travelling microscope. For improved accuracy various refinements, such as the use of a spherometer, may be adopted. Similarly, in the case of a gas bubble, d and D are measured by a travelling microscope. Measurements are usually made at room temperature but the sessile drop method for measurement of surface tension has been successfully used in the measurement of the surface tension of molten metals over a wide range of temperature.

6.10.8 Method of ripples

The rate at which ripples travel over the surface of a liquid is dependent on the surface tension and a measurement of the velocity of the ripples enables T to be determined. Consider a liquid whose surface is being traversed by transverse vertical vibrations controlled by gravity. If the waves are of small amplitude it can be assumed that each drop of liquid in the surface describes a circular path in a vertical plane. Let the velocity of the waves in a horizontal direction be c and let each drop of liquid describe a circle of radius r in an anti-clockwise direction in time t. The time taken by the waves in moving forward a distance equal to the wavelength λ is therefore also t. In *Figure 6.22(a)* at a point on a crest, A, the horizontal velocity of a drop is given by

$$u_1 = c - \frac{2\pi r}{t} \qquad \qquad \dots (6.43)$$

178

Similarly, at a point in a trough, B, the horizontal velocity of a drop is given by

$$u_2 = c + \frac{2\pi r}{t} \qquad \dots (6.44)$$

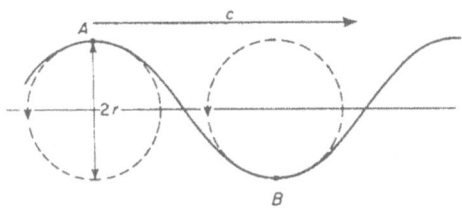

(a)

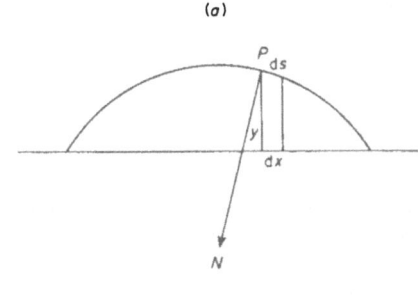

(b)

Figure 6.22. (a) Velocity of ripples; (b) effect of surface tension on ripple velocity

Assuming that the change in velocity of a drop is due to the fall in height,

$$u_2^2 = u_1^2 + 2gh \qquad \dots (6.45)$$

where $h = 2r$.

From equations (6.43) and (6.44)

$$u_2^2 - u_1^2 = \frac{8\pi rc}{t} \qquad \dots (6.46)$$

Combining equations (6.45) and (6.46)

$$4gr = \frac{8\pi rc}{t}$$

179

so that
$$c = \frac{gr}{2\pi}$$

but $\lambda = ct$ and hence

$$c = \frac{g\lambda}{2\pi c}$$

$$c = \sqrt{\frac{g\lambda}{2\pi}} \qquad \dots (6.47)$$

So far only the force of gravity has been considered, and the effect of surface tension must now be taken into account.

In the case of simple harmonic waves travelling over a liquid surface, the position of any point on the surface in a given vertical plane section, parallel to the direction of propagation, is represented by the equation

$$y = a \sin (2\pi x/\lambda + b) \qquad \dots (6.48)$$

where y is the vertical distance of the point above the undisturbed level, λ the wavelength, a the amplitude, x the distance of the point measured from some arbitrary origin and b is a constant. *Figure 6.22(b)* shows the profile curve of part of such a wave. If it is assumed that the profile curve is the same for all sections parallel to the one shown, then the wave system is cylindrical and R_2 is infinite. R_1, the radius of curvature in the plane of the diagram, is approximately equal to $1/d^2y/dx^2$, since dy/dx is very small in the case of ripples. From equation (6.48)

$$\frac{1}{R_1} = \frac{d^2y}{dx^2} = -\frac{4\pi^2 y}{\lambda^2} \qquad \dots (6.49)$$

At a point P on the surface, there is an excess pressure directed along PN equal to $-T/R_1$. Thus the force acting on a small area of length ds cm and thickness 1 cm, measured perpendicularly to the plane of the diagram, is

$$-\frac{T}{R_1} ds = \frac{4\pi^2 T y \, ds}{\lambda^2}$$

This force acts along PN, so that the force acting vertically downwards is

$$\frac{4\pi^2 T y \, dx}{\lambda^2}$$

180

The downward force on the element $y\,dx$, due to its weight, is

$$\rho g y\,dx$$

and hence the total downward force on this element is

$$g\rho y\,dx + \frac{4\pi^2 T y\,dx}{\lambda^2} = y\rho\,dx\left(g + \frac{4\pi^2 T}{\lambda^2 \rho}\right) \qquad \ldots (6.50)$$

Hence the effect of surface tension is to effectively increase g in the ratio $(1 + [4\pi^2 T/\lambda^2 \rho g])$: 1.

From equation (6.47) the velocity of waves under the influence of gravity and surface tension is given by

$$c = \sqrt{\frac{\lambda}{2\pi}\left(g + \frac{4\pi^2 T}{\lambda^2 \rho}\right)} \qquad \ldots (6.51)$$

Since $c = v\lambda$, where v is the frequency of the waves

$$T = \frac{\lambda^3 v^2 \rho}{2\pi} - \frac{g\lambda^2 \rho}{4\pi^2} \qquad \ldots (6.52)$$

Several experiments to determine surface tension by measurements on ripples have been carried out and the main problem is the accurate measurement of the wavelength. One method, originally due to Rayleigh but later improved by other workers, is as follows.

The liquid being investigated is contained in a shallow trough and its surface is cleaned by scrapers. The ripples are produced by a dipper attached to the lower prong of an electrically maintained tuning fork whose prongs are horizontal and vibrate in a vertical plane as shown in *Figure 6.23(a)*. The plane of the dipper is vertical and perpendicular to the plane of the prongs. The ripples produced by the vibration of the dipper are reflected at the ends of the trough and produce a system of stationary ripples. To enable the wave form of the ripples to be seen, and the wavelength to be measured, a viewer is attached to the lower prong of the fork. The viewer consists of two horizontal wires supported by a suitable frame. A parallel beam of light, travelling in a direction perpendicular to the plane of vibration of the prongs, is reflected by the surface of the liquid upwards past the wires to an observer, as shown in *Figure 6.23(b)*.

Due to the ripples on the liquid surface, the shadows of the wires have the wave form of the ripples, since the light is travelling in planes parallel to the crests of the real stationary ripples. A calibrated metal plate is placed in the same plane as the wires of the viewer and this is photographed with the shadow ripples. It is assumed that the shadow-ripple wavelength is the same as the wavelength of the real ripples and hence λ can be found. If the frequency of the fork is accurately determined then T can be calculated from equation (6.52).

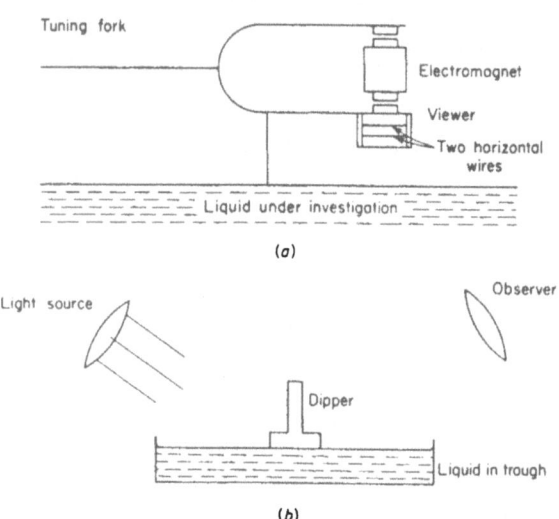

Figure 6.23. Measurement of surface tension by ripple method

6.11 VARIATION OF SURFACE TENSION WITH TEMPERATURE

The surface tension of all liquids decreases as the temperature increases until it becomes zero at the critical temperature. In 1893 van der Waals' suggested the empirical formula

$$T_\theta = A \left(1 - \frac{\theta}{\theta_c} \right)^{\frac{3}{2}} \qquad \cdots \cdots (6.53)$$

where T_θ is the surface tension at the absolute temperature θ, θ_c is the critical temperature and A is a constant for a given liquid. A modified formula, suggested by Ferguson in 1923, is

$$T_\theta = A \left(1 - \frac{\theta}{\theta_c} \right)^{n} \qquad \cdots \cdots (6.54)$$

where n is a constant for a particular liquid but is in the vicinity of 1·21. This empirical formula corresponds quite well with the

182

observed results. The experimental work also shows that the surface tension curve approaches the critical point tangentially and, in fact, the value of the surface tension is negligible at a few degrees below the actual critical temperature.

An alternative equation relating surface tension and temperature is due to Eötvös. If the liquid has a molecular weight M and density ρ, then the surface area occupied by a gramme molecule (the molar surface) is proportional to $(M/\rho)^{\frac{2}{3}}$. The surface energy in the molar surface is thus proportional to $T(M/\rho)^{\frac{2}{3}}$. Eötvös' law, subsequently modified slightly by Ramsey and Shields, states

$$T_\theta \left(\frac{M}{\rho}\right)^{\frac{2}{3}} = K(\theta_c - \theta - \delta) \qquad \ldots (6.55)$$

where T_θ is the surface tension at the absolute temperature θ, $(\theta_c - \delta)$ is the temperature where the surface tension disappears, θ_c is the critical temperature and K is a constant which has a value of approximately 2·2 for most liquids. For certain organic liquids, however, higher values of K are obtained.

6.12 RELATIONS BETWEEN SURFACE TENSION AND OTHER CONSTANTS OF A LIQUID

Macleod showed that for many liquids over a wide range of temperature, the empirical relationship below is obeyed

$$T = K(\rho - \rho_1)^4 \qquad \ldots (6.56)$$

where K is a constant and ρ and ρ_1, are the densities of the liquid and the saturated vapour, respectively. Deviations from this formula frequently occur in the vicinity of the critical temperature.

Another relationship frequently followed by many non-associated liquids, i.e. those which contain nothing but individual molecules of the liquid, is

$$T\beta^{\frac{1}{4}} = \text{constant} \qquad \ldots (6.57)$$

where β is the compressibility. This relationship was established by Richard and Mathews.

If equation (6.56) is rearranged and both sides multiplied by M (the molecular weight in grammes), then

$$\frac{MT^{\frac{1}{4}}}{\rho - \rho_1} = MC \qquad \ldots (6.58)$$

where C is a constant. If ρ_1 is neglected in comparison with ρ, then

183

$$\frac{MT^{\frac{1}{4}}}{\rho} = \text{a constant} \qquad \qquad \ldots (6.59)$$

The quantity $MT^{\frac{1}{4}}/\rho$ was called, by Sugden, the parachor P of the substance. Since M/ρ is the molecular volume V of the substance

$$P = VT^{\frac{1}{4}} \qquad \qquad \ldots (6.60)$$

Hence the parachor may be regarded as the molecular volume of a substance when its surface tension is unity.

For substances which are saturated in the chemical sense the parachor is an additive function. Isomeric compounds have parachors of the same value and the difference between the parachors of successive members of a homologous series is constant and independent of the type of compound. For instance in the paraffin series it is found that the parachor of the group CH_2, which is about 39, is the difference between the parachors of C_2H_6 and C_3H_8. This additivity law also holds for atomic parachors so that if, for instance, the parachor of n CH_2 groups is subtracted from the parachor of a paraffin C_nH_{2n+2}, the result is the parachor of molecular hydrogen.

It has been established that the parachor depends not only on the constituents of the compound but also on their mode of linkage, so that the parachor of a compound can be calculated by adding together one set of constants for the parachors of the atom in the molecule and another set of structural constants which take into account the effects of ring closure, unsaturation, etc. This has led to the parachor being applied to the determination of the structure of various chemical compounds.

6.13 SURFACE TENSION AND EVAPORATION

When a drop is evaporating under constant temperature conditions the latent heat of the vapour is partly provided by the surface energy which is lost as the surface area decreases. This means that a drop of liquid cannot be in equilibrium with an atmosphere of saturated vapour round it since, as the drop becomes smaller, the stage is reached where the loss in surface energy is sufficient to supply all the latent heat of vaporization. Conversely, for the formation of a drop, an initial nucleus is necessary and for condensation to take place supersaturation of the vapour is necessary, to a degree depending on the size of the nuclei available. Usually these nuclei are dust particles and drop formation requires only a small degree of supersaturation. In 1897 C. T. R. Wilson showed that electrically charged molecules also act as nuclei for the condensation of super-saturated vapour and this discovery led to the development of the cloud chamber for the detection of nuclear radiation.

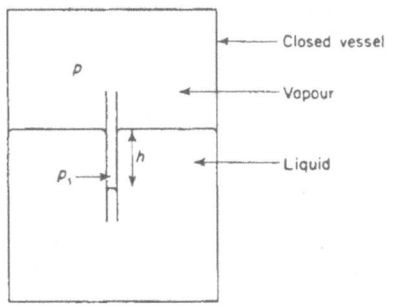

Figure 6.24. Vapour pressure over a curved surface

The effect of the curvature of a liquid surface on the equilibrium between the liquid and its vapour, may be shown as follows. *Figure 6.24* shows a closed vessel containing the liquid and its vapour. A capillary tube of a material not wetted by the liquid is placed vertically in the liquid as shown. If p is the saturation vapour pressure at the horizontal liquid surface and p_1 the equilibrium pressure above the curved surface of radius r, then

$$p_1 - p = \int g\sigma \, dh \qquad \qquad \dots (6.61)$$

where σ is the vapour density and h is the depth of the meniscus below the horizontal liquid surface. Now the excess pressure inside the curved surface is $2T/r$, hence

$$p + \rho gh - \frac{2T}{r} = p_1 \qquad \qquad \dots (6.62)$$

where ρ is the liquid density.
    Thus, from equations (6.61) and (6.62)

$$\int g\sigma \, dh = \rho gh - \frac{2T}{r}$$

or $$\frac{2T}{r} = \int g(\rho - \sigma) \, dh \qquad \qquad \dots (6.63)$$

But, $\sigma = p/R\theta$ and $dp = g\sigma \, dh$

therefore
$$\frac{2T}{r} = \int_p^{p_1} \frac{\rho - \sigma}{\sigma} \cdot dp \simeq \int_p^{p_1} \frac{\rho}{\sigma} \, dp$$

therefore
$$\frac{2T}{r} = \rho R \theta \int_p^{p_1} \frac{dp}{p} = \rho R \theta \log \frac{p_1}{p}$$

therefore
$$\log \frac{p_1}{p} = \frac{2T}{\rho R \theta r} \qquad \qquad \dots (6.64)$$

Equation (6.64) thus gives the ratio of the vapour pressures for equilibrium with the curved surface of radius r and a flat surface. For a given degree of supersaturation, i.e. a certain value of p_1, drops of radius less than r will evaporate, while those of greater size will grow.

6.14 SURFACE FILMS

If a drop of a non-volatile substance is placed on the surface of a liquid in which it is insoluble, then the substance may remain as a compact drop on the liquid surface, or it may spread out over the liquid surface. The thickness of the film formed on the liquid surface, depends on the surface area to be covered and if this is large enough it is possible to have a surface film only one molecule thick. This is referred to as a 'monomolecular' surface film. If the surface area is not of sufficient extent for the whole of the substance to exist as a monomolecular film, it is found that a monomolecular film is formed over the whole surface and this is interspersed with small droplets of much greater thickness.

The properties of surface films may conveniently be described by the use of the concept of surface pressure, or the differential surface tension. Since the films are nearly all monomolecular, each molecule moves about in two dimensions colliding with other molecules and with the surface boundaries. The momentum per metre per second given to the boundaries can be regarded as the force per metre length exerted by the film on the boundary, and this force is the surface pressure P, of the film. It is also equal to the difference of the surface tensions of the pure liquid T_1, and of the liquid covered by the surface film T_2, so that

$$P = T_1 - T_2 \qquad \qquad \dots (6.65)$$

The surface pressure can be measured by a surface tension balance. Basically this consists of a liquid contained in a long rectangular shallow dish, the liquid surface being divided by a floating barrier. The barrier is attached to a balance, arranged so that the horizontal force on one side of the barrier can be measured. If a surface film is placed on the liquid surface on one side of the barrier, the force exerted on the barrier can be measured. If the area of the surface film can be controlled, the variation of the surface pressure with the area of the film may be investigated.

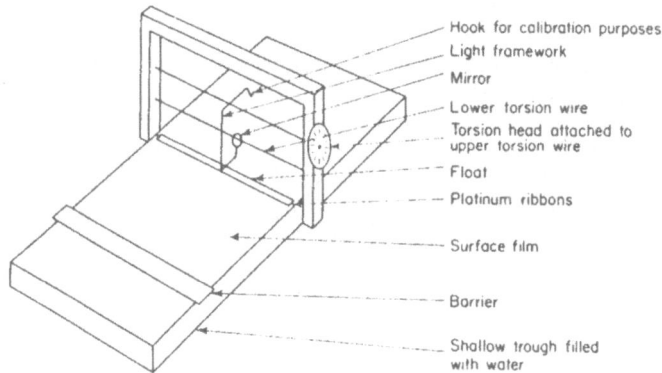

Figure 6.25. Adam's surface tension balance

Early work on the measurement of surface pressure was done by Langmuir who devised a very sensitive balance. His technique was later improved by Adam, and Adam's surface tension balance was used to investigate the behaviour of various films on water, as shown in *Figure 6.25*. It consists of a brass trough, coated with paraffin wax, which is completely filled with clean water. The surface film is confined to the area between the barrier and the float and its area can be varied by adjustment of the barrier. The float, upon which the pressure of the film is exerted, consists of a vertical sheet of waxed copper foil which extends almost the whole width of the trough and which dips into the water. The gaps at each end of the float are closed by thin platinum ribbons, thus ensuring that the film cannot pass beyond the float. The vertical framework has two horizontal torsion wires stretched across it, the upper one being attached to the torsion head. The lower wire carries a mirror and is connected to the float.

Thus, any movement of the float causes a rotation of the mirror and hence of a beam of light reflected from it. The upper torsion wire is attached to the float *via* the light framework so that any rotation of the mirror can be counterbalanced by rotation of the torsion head with consequent movement of a pointer over a scale.

187

Once the balance is calibrated the surface pressure can be found. Calibration is carried out by suspending known weights on the small hook on the light framework. A knowledge of the dimensions of the framework thus enables the force, in dynes, on the centre of the float to be determined for a known weight on the hook, and hence the torsion head may be calibrated.

In an experiment the water surface is first thoroughly cleaned by drawing over it a paraffin wax-coated glass barrier. The torsion head is adjusted to the zero position and the light reflected from the mirror is also adjusted to the zero on the scale. The surface film substance, usually dissolved in a solvent immiscible with water, is placed on the water surface, the area of the film formed being determined by the position of the barrier. The spot of light is kept on the zero mark by adjustment of the torsion head and hence the surface pressure may be measured. Forces of the order of 10^{-7} N may be measured by this apparatus.

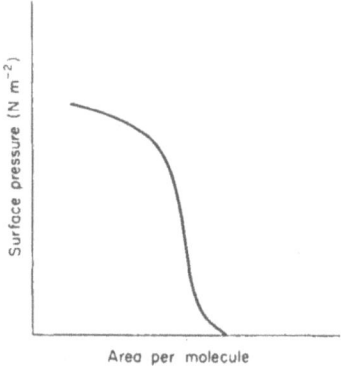

Figure 6.26. Typical result for surface pressure of a thin film

The results of such experiments are usually expressed graphically when surface pressure is plotted against the surface area of one molecule of the film substance. This area may be calculated from a knowledge of the mass of substance occupying the known area, its molecular weight, and Avogadro's number. *Figure 6.26* shows a typical curve. It is found that when the value of the surface area of one molecule is larger than a certain critical value, there is no measurable surface pressure. As the area is decreased, the value of surface pressure increases gradually until a stage is reached where it remains almost constant. In the case of fatty acids it is found that the area occupied by one molecule when the surface pressure

188

becomes measurable is always the same and this indicates that the hydrocarbon chains in the series of fatty acids must be steeply orientated to the water surface.

A tremendous amount of work has been carried out on the properties of surface films, and a fuller account of this work may be found in various specialized works such as 'Interfacial Phenomena' by J. T. Davies and E. K. Rideal.

7

VISCOSITY

7.1 INTRODUCTION

A FLUID is normally defined as a substance which is incapable of sustaining a shearing stress. However, this definition is only applicable when the fluid is at rest. If relative motion takes place a measurable resistance is encountered and the fluid is said to exhibit viscosity. Consider a fluid flowing without turbulence over a fixed surface AB (*Figure 7.1*). Experimentally, it is found that a layer of the fluid at D, at a distance $x + dx$ from AB, moves with a velocity $u + du$ greater than that of a layer of the fluid at C, at a distance x from AB, which moves with velocity u. The velocity gradient between the layers C and D is thus du/dx.

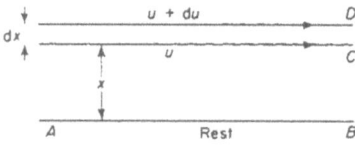

Figure 7.1. Law of viscous flow

Newton assumed that for a fluid moving in parallel layers, the shearing stress at any point is directly proportional to the velocity gradient and thus formulated his law of viscous flow

$$\frac{F}{A} \propto \frac{du}{dx}$$

or
$$\frac{F}{A} = \eta \frac{du}{dx} \qquad \ldots\ldots (7.1)$$

where F is the tangential viscous force between two layers of area A, at a distance dx apart, moving with relative velocity du. η is called the coefficient of viscosity of the fluid.

This law is valid only when the fluid is moving with streamline motion. It is not applicable when turbulence occurs. From equation

190

(7.1) it can be seen that $\eta = F/A \, du/dx$ and the dimensions of η are $ML^{-1}T^{-1}$. In the SI system the unit is thus Nsm.^{-2} Alternatively the centipoise, equal to $10^{-3} \, \text{Nsm}^{-2}$, may be used. An alternative expression to viscosity is the so-called kinematic viscosity of a fluid. This is equal to η/ρ where ρ is the density of the fluid concerned and the unit in the SI system is the centistoke equal to $10^{-6} \, \text{m}^2\text{s}^{-1}$.

7.2 THE STEADY FLOW OF AN INCOMPRESSIBLE LIQUID THROUGH A NARROW HORIZONTAL TUBE

Consider a liquid flowing through a narrow horizontal tube of radius a and length l, under a pressure difference P between the ends of the tube. Assume the liquid moves with streamline motion parallel to the axis of the tube and that the liquid in contact with the walls of the tube is at rest.

When steady conditions are attained let the velocity of the liquid at a distance r from the axis of the tube be u and let the velocity gradient be du/dr. This viscous drag per unit area is thus $\eta \, du/dr$ and this acts over all the surface area of the inner cylinder of liquid in a direction opposed to the pressure gradient down the tube. The total viscous drag on this cylinder of liquid is thus $\eta \, du/dr \,.\, 2\pi rl$. The force tending to accelerate the liquid cylinder is $P \,.\, \pi r^2$ and thus, when steady conditions exist, this force must be balanced by the viscous drag, i.e.

$$P\pi r^2 = -\eta \frac{du}{dr} \,.\, 2\pi rl$$

$$-r \, dr = \frac{2\eta l}{P} \,.\, du \qquad \qquad \dots (7.2)$$

At the walls of the tube $r = a$ and $u = 0$. Integrating equation (7.2) between the limits $r = a$ to $r = r$ gives

$$a^2 - r^2 = \frac{4\eta l}{P} \,.\, u$$

i.e. $$u = \frac{P}{4\eta l}(a^2 - r^2) \qquad \qquad \dots (7.3)$$

The volume of liquid which flows through the tube per second between the radii r and $r + dr$ is given by

$$dV = 2\pi r \, dr \,.\, u = \frac{P\pi}{2\eta l}(a^2 - r^2)r \, dr$$

191

and hence, the total volume of liquid flowing through the tube per second may be obtained by integration and is given by

$$V = \int_0^a \frac{P\pi}{2\eta l}(a^2 - r^2)r\, dr = \frac{P\pi a^4}{8\eta l} \qquad \dots (7.4)$$

This expression relating the volume of liquid flowing through a tube per second, V, to the pressure difference, P, the coefficient of viscosity η, and the length l and radius, a, of the tube, is known as Poiseuille's formula.

7.3 CORRECTIONS TO POISEUILLE'S FORMULA

In the derivation of equation (7.4) two factors were neglected and for accurate work these must be taken into account. First accelerations occur near the inlet of the tube and these do not decrease to zero until an appreciable part of the tube has been traversed. This error must thus be corrected for and this is done by increasing the value of l by a quantity $K_1 a$ where K_1 is a constant approximately equal to 1·64. Secondly the pressure difference between the ends of the tube is partly used in giving kinetic energy to the liquid and is not wholly used in overcoming viscous forces. If the effective pressure difference which overcomes the viscosity is P_1 then in 1 s the work done against viscous forces is $P_1 V$ while the kinetic energy given to the liquid is

$$\int_0^a \tfrac{1}{2}\rho 2\pi r\, dr \,.\, u \,.\, u^2 = \pi\rho\left(\frac{P_1}{4\eta l}\right)^3 \int_0^a r(a^2 - r^2)^3\, dr$$

$$= \pi\rho\left(\frac{P_1}{4\eta l}\right)^3 \frac{a^8}{8}$$

$$= \left(\frac{P_1\pi a^4}{8\eta l}\right)^3 \frac{\rho}{\pi^2 a^4} \qquad \dots (7.5)$$

The total loss of energy is therefore

$$P_1 V + \frac{V^3\rho}{\pi^2 a^4}$$

Since this must be equal to PV

$$P_1 = P - \frac{V^2\rho}{\pi^2 a^4} \qquad \dots (7.6)$$

192

This correction has been tested experimentally by Hagenbach and other workers and has been found to be approximately true. The correct expression is

$$P_1 = P - \frac{K_2 V^2 \rho}{\pi^2 a^4} \qquad \qquad \dots (7.7)$$

where K_2 is a constant whose value depends on the apparatus used. Its value must be obtained by calibration and it is found to be almost always nearly unity.

The application of these two corrections to Poiseuille's equation results in equation (7.4) becoming

$$V = \frac{\left(P - \dfrac{K_2 V^2 \rho}{\pi^2 a^4}\right) \pi a^4}{8\eta(l + K_1 a)}$$

i.e. $$\eta = \frac{P\pi a^4}{8V(l + 1 \cdot 64a)} - \frac{V\rho K_2}{8\pi(l + 1 \cdot 64a)} \qquad \dots (7.8)$$

7.4 CRITICAL VELOCITY

In the derivation of Poiseuille's equation it was assumed that the liquid moved with streamline motion parallel to the axis of the tube. It is important to note that Poiseuille's equation is only valid as long as this condition is satisfied. For tubes of small radius and for low velocities of the liquid this condition is indeed fulfilled. If the velocity of the liquid is increased beyond a certain value however, so that turbulence occurs, then Poiseuille's equation is no longer valid. The liquid velocity at which turbulence begins is called the critical velocity, v_c. Reynolds showed experimentally that for a liquid of density ρ and coefficient of viscosity η moving through a tube of radius a, the critical velocity is given by

$$v_c = \frac{K\eta}{\rho a} \qquad \qquad \dots (7.9)$$

where K is a constant known as Reynolds' number and is approximately equal to 1000. This formula can be simply derived by the method of dimensional analysis. If it is assumed that the critical velocity is dependent on the factors, ρ, η and a, then

$$v_c = K\eta^x \rho^y a^z \qquad \qquad \dots (7.10)$$

193

Putting in the appropriate dimensions

$$[LT^{-1}] = [ML^{-1}T^{-1}]^x [ML^{-3}]^y [L]^z$$

Equating indices gives

For L: $\qquad\qquad 1 = z - 3y - x$

For M: $\qquad\qquad 0 = x + y$

For T: $\qquad\qquad -1 = -x$

Hence $\qquad\qquad\qquad x = 1$

$\qquad\qquad\qquad\qquad y = -1$

and $\qquad\qquad\qquad z = -1$

Substituting these values in equation (7.10) gives

$$v_c = \frac{K\eta}{\rho a}$$

At low liquid velocities the volume of liquid flowing through the tube is given by Poiseuille's equation. When the critical velocity is exceeded the volume of liquid becomes independent of the liquid viscosity and becomes mainly dependent on the density. For turbulent flow it is found that the volume of liquid flowing through the tube is almost proportional to the square root of the pressure. In turbulent conditions the pressure difference is used up in over-coming the turbulent motion and in communicating kinetic energy to the liquid.

7.5 DETERMINATION OF THE COEFFICIENT OF VISCOSITY OF A LIQUID BY FLOW THROUGH A CAPILLARY TUBE

The apparatus is illustrated in *Figure 7.2*. It consists of a piece of capillary tubing about 0·5 m long with a uniform bore of diameter 2–3 mm. Two small holes are bored in the tube about 0·3 m apart. T-pieces are fitted over the ends of the capillary tube as shown and are adjusted so that the centres of the T-pieces lie over the two holes bored in the capillary tube. The joints between the tubes are made leakproof, usually by sealing with wax. The T-tubes are connected to the tubes forming a manometer and thus the pressure difference over the length of the capillary tubing can be measured directly.

The liquid is fed into the capillary tubing from a constant head device as shown. The volume of liquid passing through the tubing in a fixed period of time is measured and, after the experiment, the diameter of the tubing is accurately determined by weighing the length of a mercury thread which fills a measured length of the tube. Hence in Poiseuille's equation, all the quantities are known except η which can thus be calculated.

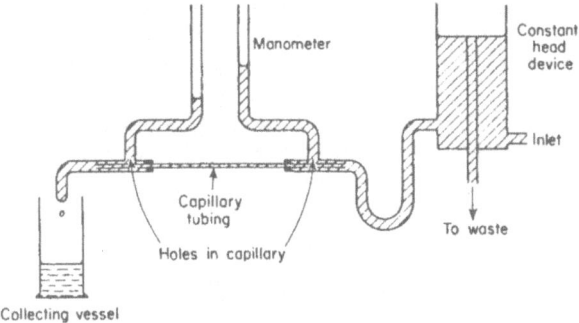

Figure 7.2. Viscosity of a liquid by flow method

7.6 VARIATION OF VISCOSITY OF A LIQUID WITH TEMPERATURE

The variation of the viscosity of a liquid with temperature can be conveniently examined with the apparatus shown in *Figure 7.3*. It consists of a wide vessel, containing the liquid at a known temperature, provided with a siphon of glass tubing connected to a

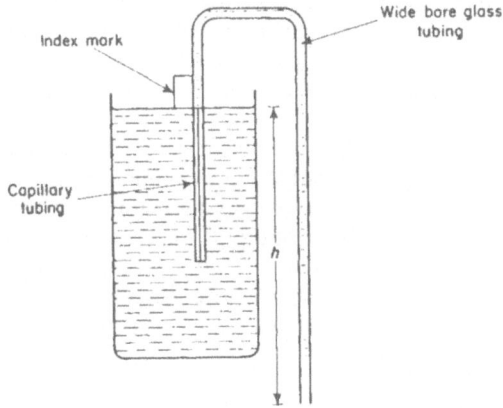

Figure 7.3. Variation of viscosity of a liquid with temperature

length of capillary tubing. The siphon is provided with an index point which is adjusted to touch the surface of the liquid. The liquid is sucked over and as the liquid level in the vessel falls, the siphon

195

is lowered in order to just keep the index point touching the liquid surface. The volume of liquid flowing through the tubing in a known time is determined and the length and diameter of the capillary tubing determined in the usual way. The pressure difference P is equal to $\rho g h$ where ρ is the liquid density and h is the vertical distance between the top of the capillary tubing and the bottom of the wide bore glass tubing. It is seen from the diagram that the column of liquid in the capillary tube is supported by the upthrust of the surrounding liquid, thus leaving the column of $\rho g h$ to provide the pressure necessary to overcome the viscous forces. For experiments at various temperatures the vessel can be provided with a heating coil and suitable lagging.

7.7 TYPES OF VISCOMETER

7.7.1 The rotation viscometer

Both experiments described so far for the determination of the coefficient of viscosity of a liquid were based on the flow of the liquid through a capillary tube. Several other methods for measuring viscosity are available. These are based on observations of the motion of a solid body moving in the liquid and some of the more important of these methods are discussed in the next few sections.

The rotation viscometer consists of a vertical cylinder which is made to rotate uniformly within a coaxial cylinder of known radius, the liquid under investigation being contained in the annular space between the cylinders. Since only relative rotation is involved, theoretically it is immaterial whether the inner cylinder is rotated and the outer cylinder kept stationary or vice versa. The theory of the method is as follows.

Let the inner and outer cylinders have radii a and b respectively and let the liquid cover a length l of the inner cylinder. Let the inner cylinder be at rest and the outer cylinder rotate with angular velocity Ω. Thus the innermost layer of liquid will be at rest while the outermost layer has a speed of $b\Omega$.

Consider the forces acting over the side of the liquid cylinder of radius r whose angular velocity of rotation is ω. *Figure 7.4.* The velocity gradient at this point is

$$\frac{\mathrm{d}}{\mathrm{d}r}(r\omega) = \omega + r\frac{\mathrm{d}\omega}{\mathrm{d}r} \qquad \ldots (7.11)$$

The first term on the right-hand side of this equation represents the angular motion the layer would have if no viscous slip occurred and thus only the second term is responsible for the viscous effects.

196

The layer of liquid outside the cylinder of radius r and length l, exerts on this cylinder a force $2\pi r l\eta . r(d\omega/dr)$ and the moment of this about the axis is

$$\tau_1 = 2\pi l\eta r^3 \frac{d\omega}{dr} \qquad \dots (7.12)$$

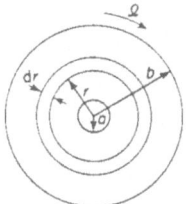

Figure 7.4. Example

In the steady state this torque must be equal to that exerted on the inner cylinder. Integration of equation (7.12) between the limits a and b gives

$$\tau_1\left[\frac{1}{a^2} - \frac{1}{b^2}\right] = 4\pi\eta l\Omega \qquad \dots (7.13)$$

τ_1 is the torque acting over the side of the cylinder but there is also a torque τ_2 due to viscosity acting over the base of the inner cylinder. The magnitude of τ_2 depends on a and b and the distance between the bases of the two cylinders. If these quantities remain fixed while the length of the inner cylinder acted on by the liquid is varied then, while τ_1 varies; τ_2 remains the same. Hence the torque τ_2 can be eliminated by measurement with two different lengths of cylinder immersed. If τ_2 is put equal to some function $f(a, b)$, then for two different lengths l_1 and l_2

$$\Gamma_1 = \tau_1 + \tau_2 = \frac{4\pi\eta\Omega a^2 b^2}{b^2 - a^2} l_1 + f(a, b)$$

and

$$\Gamma_2 = \tau_1' + \tau_2 = \frac{4\pi\eta\Omega a^2 b^2}{b^2 - a^2} l_2 + f(a, b)$$

where Γ_1 and Γ_2 are the total torques acting on the cylinder.

Hence

$$\Gamma_1 - \Gamma_2 = \frac{4\pi\eta\Omega a^2 b^2}{b^2 - a^2} (l_1 - l_2) \qquad \dots (7.14)$$

197

H

Equation (7.14) applies equally to liquids and gases. The validity of application of the equation is dependent on conditions of streamline flow. At high values of Ω, when turbulence sets in, the relationship between τ_1 and Ω is no longer linear and in fact τ_1 is approximately proportional to Ω^2.

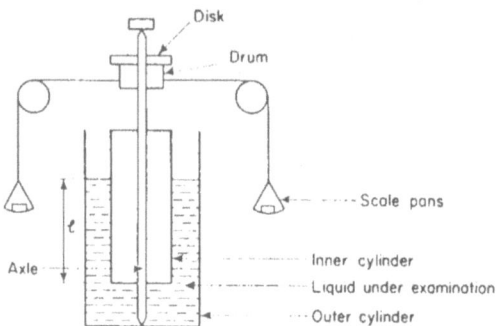

Figure 7.5. Searle's rotation viscometer

Figure 7.5 shows the rotation viscometer developed by Searle for use with liquids. It consists of a solid inner cylinder, of radius a, fixed to an axle pivoted freely at both ends. A disk is attached to the upper end of the axle in order to facilitate measurement of the period of rotation of the cylinder. The cylinder is mounted coaxially within a fixed cylinder, of radius b, and the liquid under examination is placed between the two cylinders so that it covers a length l of the inner cylinder. The inner cylinder is caused to rotate by the couple provided by the weights in the two scale pans, the couple being transmitted by cords over frictionless pulleys to a drum. When the inner cylinder is rotating steadily then, from equation (7.13)

$$\eta = \frac{\tau_1}{4\pi l \Omega}\left(\frac{1}{a^2} - \frac{1}{b^2}\right)$$

If the diameter of the drum is d and the weight in each scale pan is mg, then the couple $\tau_1 = mgd$. If the period of rotation of the cylinder is T then since $T = 2\pi/\Omega$

$$\eta = \frac{gd}{8\pi^2}\frac{(b^2 - a^2)}{(a^2 b^2)}\frac{mT}{l} \qquad \dots (7.15)$$

Thus, a graph of mT against l is a straight line. Such a graph intersects the l axis on the negative side of the origin thus giving the 'end effect' correction factor which must be added to the length l in equation (7.15). Since all the quantities in equation (7.15) can be measured, η may be calculated from the slope of the graph of mT against l.

A typical apparatus has the following dimensions, $a = 2$ cm, $b = 3$ cm, $l = 7$ cm, $d = 2$ cm and $m = 50–100$ g. These values result in a period of about 25 s for liquids of large viscosity such as syrup.

7.7.2 Stoke's falling body viscometer

Stoke's law, which was derived from hydrodynamical considerations for a perfectly homogeneous continuous fluid of infinite extent, states that

$$F = 6\pi\eta au \qquad \ldots (7.16)$$

where F is the viscous retarding force exerted on a sphere, of radius a, moving with uniform velocity u, through a fluid whose coefficient of viscosity is η.

It was shown by Newton that when a body is acted on by a constant force and is also subjected to a resistance proportional to its velocity, then ultimately, when the force resisting its motion is equal and opposite to the constant force causing the motion, the body attains a constant or terminal velocity. Thus, when a sphere is allowed to fall under gravity in a viscous fluid, when it has attained its terminal velocity, the viscous retarding force is equal to the force causing the motion of the sphere, i.e. its weight. Hence

$$6\pi\eta au = \tfrac{4}{3}\pi a^3(\rho - \sigma)g \qquad \ldots (7.17)$$

where ρ is the density of the material of the sphere and σ is the density of the liquid. Equation (7.17) affords a convenient method for the determination of the coefficient of viscosity of a liquid where the liquid is available in an appreciable quantity. Basically, the experiment involves allowing metal spheres of known radius and density to fall in a vertical glass tube filled with the liquid under investigation. Measurement of the transit time between two fixed marks on the side of the tube gives the terminal velocity and thus the only unknown in equation (7.16) is η, the coefficient of viscosity.

The apparatus is shown in *Figure 7.6*. It consists of a large glass cylinder, with a length of the order of 1 m and a diameter of 10 cm, filled with the liquid whose coefficient of viscosity is required. Two fixed marks are made by winding black cotton round the cylinder at distances roughly one-third and two-thirds of the distance from the top of the vessel respectively. It is assumed that the sphere has attained its terminal velocity in the first third of the liquid and the subsequent terminal velocity is determined by measurement of the transit time between the fixed points, a known distance apart.

The spheres are steel ball bearings of various diameters, of the order of 1–4 mm. Their diameter is carefully measured with a micrometer screw gauge and they are thoroughly wetted with the liquid under investigation. Each sphere is subsequently fed, in turn,

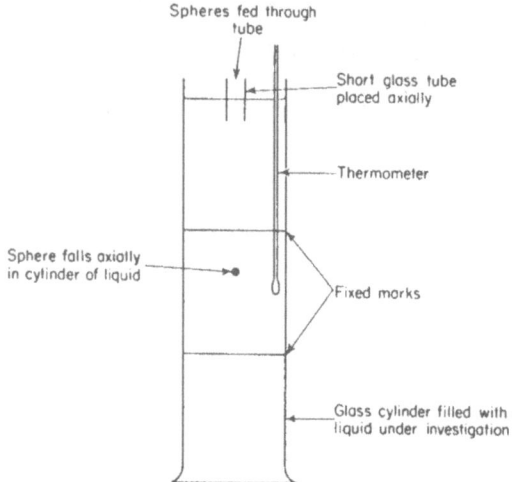

Figure 7.6. Stoke's falling body viscometer

through a short vertical glass tube placed centrally as shown, thereby ensuring that they fall axially through the liquid. The transit time is carefully noted for each sphere, several different measurements being made for each different diameter. Now equation (7.17) can be rearranged in the form

$$\eta = \frac{2}{9} a^2 g (\rho - \sigma) \frac{t}{s} \qquad \dots (7.18)$$

where s is the distance between the fixed marks and t is the transit time. Hence, for a given liquid at constant temperature, $a^2 t$ is constant and the graph of t against $1/a^2$, should be a straight line. The value of η, for the liquid being investigated, may then be obtained from the slope of the line. The temperature of the liquid must be maintained constant since viscosity varies with temperature. An improvement on the experiment described above is to place the cylinder inside a constant temperature bath.

Equation (7.16) was derived by Stokes for a sphere falling in a continuous fluid of infinite extent. In the experiment described, therefore, corrections must be made for the boundary conditions appertaining at the walls and base of the cylinder containing the liquid. Ladenburg showed that to correct for the wall effect, the true velocity of the sphere is given by

$$u_\infty = u\left(1 + 2\cdot4\,\frac{a}{R}\right) \qquad \ldots\,(7.19)$$

where u is the observed velocity and R is the radius of cross-section of the cylinder containing the fluid. Similarly, to correct for the effect of the base of the cylinder, the following formula is used

$$u_\infty = u\left(1 + 3\cdot3\,\frac{a}{h}\right) \qquad \ldots\,(7.20)$$

where h is the total height of the liquid. Hence, incorporating these two corrections into equation (7.18) gives

$$\eta = \frac{2}{9}\frac{(\rho - \sigma)\,ga^2}{u[1 + 2\cdot4(a/R)]\,[1 + 3\cdot3(a/h)]} \qquad \ldots\,(7.21)$$

7.7.3 Revolving disk viscometer

This method of measuring coefficients of viscosity is based on the couple which acts on a suspended disk when it is in contact with a rotating fluid. The apparatus is shown in *Figure 7.7* and consists of

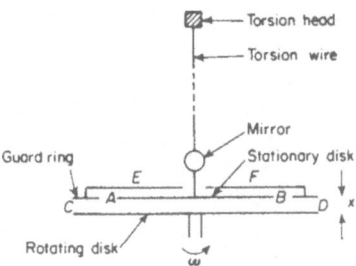

Figure 7.7. Revolving disk viscometer

a thin circular disk, of diameter about 40 cm, mounted so that it can be rotated with constant angular velocity about a vertical axis through its centre. A few millimetres above it another disk is suspended by means of a torsion wire, so that the centre of the upper disk is vertically above that of the lower disk, the planes of the disks being parallel. When the lower disk is rotated at a constant rate of about 1 rev/s, a couple is exerted on the upper disk. This causes it to move through an angle θ until the couple due to the viscous drag is equal to the restoring couple due to the twist in the torsion wire.

If it is assumed that the angular velocity of any layer of liquid between the disks is a linear function of the distance of the layer from the upper disk then, if the angular velocity of the rotating disk is ω, and the distance between the disks is x, the vertical velocity gradient at a distance r from the axis of rotation will be equal to $r\omega/x$. Hence the torque acting on the portion of the upper ring between radii r and $r + dr$ is

$$d\tau = 2\pi r dr \cdot \eta \cdot \frac{r\omega}{x} \cdot r$$

$$= 2\pi r^3 \frac{\eta\omega}{x} dr$$

Thus the total torque is given by

$$\tau = 2\pi\eta \frac{\omega}{x} \int_0^a r^3 dr$$

therefore

$$\tau = \frac{\pi\eta\omega a^4}{2x} \qquad \qquad \dots (7.22)$$

where a is the radius of the upper disk. This formula is only approximate, since the average velocity gradient $r\omega/x$ is not necessarily that at the surface of the disk and also, at the edge of the disk, the distribution of streamlines is complex. The error arising from this second effect may be reduced by mounting a guard ring round the suspended disk as shown in the diagram. The upper disk AB is arranged to almost fill a circular aperture in a larger disk CD, and in this way, end effects are eliminated. The cap EF, resting on CD, serves to protect the upper surface of the disk AB from viscous drag. The angular deflection of the upper disk, θ, is determined in the usual way by means of the mirror attached rigidly to the disk. If the couple per unit twist in the torsion wire is c, then, when the disk is in equilibrium

$$\tau = c\theta$$

and

$$\eta = \frac{2xc\theta}{\pi\omega a^4} \qquad \qquad \dots (7.23)$$

7.8 COMMERCIAL VISCOMETERS

In industry, when a knowledge of the coefficient of viscosity of a liquid is required, it is not usual to carry out an absolute determination by one of the methods described above. Instead, much simpler instruments are used which enable relative measurements of viscosity to be made. Once calibrated, these instruments can, of course, provide an absolute measure of viscosity.

7.8.1 Ostwald viscometer

The most widely used commercial viscometer is that due to Ostwald. This is shown in *Figure 7.8*. Basically it consists of a U-tube with a capillary tube as one side. A constant volume of liquid is poured into the right-hand limb of the tube and this is sucked round until the part of the tube from A to C is filled with the liquid. The liquid is then allowed to flow and the time taken for the liquid level to fall from A to B is determined. The theory of the viscometer is as follows.

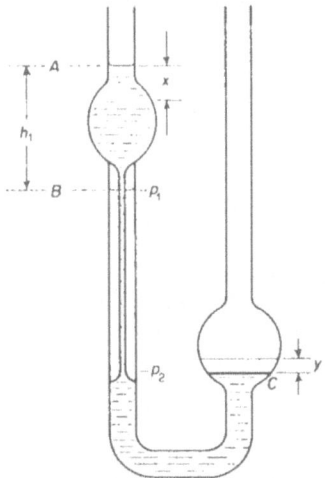

Figure 7.8. *Ostwald viscometer*

Let the cross-section of the upper bulb be a function of x only, so that $A = f(x)$ and similarly let the cross-section of the lower bulb be a function of y only so that $C = f'(y)$. Here, x is the vertical distance which the liquid meniscus falls in the upper bulb in time t, while y is the distance through which the meniscus rises in the lower bulb in the same time. If the volume of liquid which has passed through the vertical capillary tube in the time t is V, then, from equation (7.4), taking into account the fact that the capillary tube is vertical,

203

$$V = \left[\frac{\pi a^4}{8\eta l}(p_1 - p_2 + \rho g l)\right] t \qquad \dots (7.24)$$

where l is the length of capillary tubing and ρ is the liquid density. From equation (7.24)

$$dV = \frac{\pi a^4}{8\eta l}(p_1 - p_2 + \rho g l)\, dt$$

If the fall in the upper bulb in time dt is dx and the corresponding rise in the lower bulb is dy, then

$$dV = A\,dx = C\,dy$$

Hence

$$A\,dx = \frac{\pi a^4 \rho g}{8\eta l}(h_1 - x - y + l)\, dt \qquad \dots (7.25)$$

This assumes that the initial level of the liquid in the lower bulb is in the same horizontal plane as the lower end of the capillary tube, thus

$$A\,dx = \frac{k\rho}{\eta}(k_1 - x - y)$$

where k and k_1 are constants.

Since $V = \int_0^t dV = \int_0^x f(x)\,dx = \int_0^y f'(y)\,dy$, and, y is a function of x only, then

$$A\,dx = \frac{k\rho}{\eta} f(x)\, dt \qquad \dots (7.26)$$

where $f(x)$ is written for $(k_1 - x - y)$.

If T is the time for the liquid level to fall from the fixed marks at A and B so that $x = X$ say, then, from equation (7.26)

$$\int_0^x \frac{A\,dx}{f(x)} = \frac{k\rho T}{\eta}$$

For a given instrument at constant temperature $x \int_0^2 A\,dx/f(x)$ is constant and hence $k\rho T/\eta$ is constant. Thus if equal volumes of two liquids, whose viscosities are to be compared, are used in turn in the viscometer, if the times of fall are T_1 and T_2

$$\frac{\eta_1}{\eta_2} = \frac{\rho_1 T_1}{\rho_2 T_2} \qquad \dots (7.27)$$

Remember that the kinematic viscosity of a liquid is defined as the ratio η/ρ and hence the ratio of the times of fall in the Ostwald viscometer is directly equal to the ratio of the kinematic viscosities of the liquids. The Ostwald viscometer is widely used in industry and it is frequently used within a constant temperature bath for providing accurate measurements of viscosity over a wide range of temperature.

7.8.2 Redwood viscometer

Another viscometer widely used in industry for comparative measurements of viscosity is the Redwood viscometer. The main feature of this instrument is a small agate plug through which there is a capillary tube. The plug is fitted into the bottom of a cylindrical metal vessel which can be provided with a heating coil and stirrer to bring the liquid to the temperature at which the viscosity is to be measured. The instrument is initially calibrated after which viscosity determinations can be carried out simply by timing the rate of fall of a fixed volume of liquid between two fixed fiduciary marks on the cylinder.

7.8.3 Mitchell cup and ball viscometer

A useful viscometer, if only small amounts of liquid are available, is the Mitchell cup and ball viscometer. This is illustrated in *Figure 7.9* and consists of a hemispherical cup into which fits a steel ball.

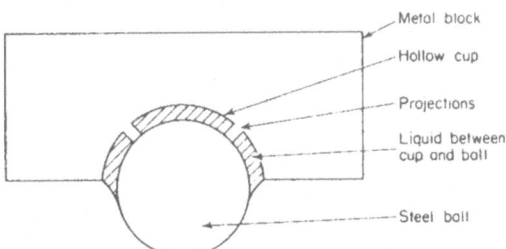

Figure 7.9. The Mitchell cup and ball

Three projections prevent the cup and ball from touching. The liquid is placed in the cup, the ball put in place, then the instrument is inverted so that the ball would drop were it not for the liquid. Ultimately, depending on the liquid viscosity, the ball falls and the length of time the ball is held in the cup affords an approximate measure of viscosity once the instrument is calibrated. The Mitchell

205

H*

cup and ball is useful for approximate determinations of the viscosity of heavy oils.

7.9 EFFECT OF TEMPERATURE AND PRESSURE ON THE VISCOSITY OF LIQUIDS

The viscosity of liquids is extremely dependent on temperature and it is found that an increase in temperature produces a marked decrease in viscosity. Hence, any measurement of viscosity must always be done at a constant temperature and this temperature must be quoted with the value of the viscosity. At the present time no simple formula has been derived connecting the variation of viscosity and temperature, which is valid for all liquids. Various empirical formulae have been suggested, that due to Slotte being

$$\eta_t = \eta_0(1 + at + bt^2)^{-1} \qquad \ldots (7.28)$$

where η_t is the viscosity at $t°C$, a and b being constants. This formula is not, however, in very good agreement with experimental results. A modified formula

$$\eta_t = A(1 + Bt)^c \qquad \ldots (7.29)$$

where A, B and C are constants, agrees fairly well with experimental observations on pure liquids but does not agree with the results obtained for oils.

Andrade considered a liquid to consist of molecules vibrating under the influence of local forces about equilibrium positions which, instead of being fixed as in a solid, are slowly displaced with time. On the basis of this theory he derived the following formula connecting viscosity and temperature

$$\eta\rho^{-\frac{1}{3}} = A \exp C\rho/T \qquad \ldots (7.30)$$

where ρ is the liquid density, T the absolute temperature and A and C are constants. This formula is in good agreement with experimental results for most liquids. It does not, however, agree with the results for water and certain tertiary alcohols.

The work of Bridgman has established that, with the exception of water, all liquids behave similarly as the pressure is increased. He showed that the viscosity of liquids increases with pressure at a rapidly increasing rate, although the actual quantitative values of viscosity show a large variation from one liquid to another. Andrade derived a formula connecting viscosity and pressure and this is in good agreement with the observed results up to pressures of about 2000 atm.

Water behaves in an unusual manner. Between $0°$ and $10°C$ it exhibits a minimum viscosity at about 1000 atm, but at a higher temperature it shows a steady increase in viscosity with increase in pressure.

7.10 FLOW OF A COMPRESSIBLE FLUID THROUGH A NARROW TUBE

The derivation of Poiseuille's equation, equation (7.4), was made assuming that the liquid was incompressible so that the volume crossing any section of the tube was constant. This assumption is true in the case of liquids where the density is independent of the pressure, but it is not true in the case of gases. With gases it is the mass per unit time crossing any section of the tube which is constant. If ρ is the density of the gas at the point considered, A the cross-sectional area of the tube and u the velocity of gas flow, then $\rho A u$ is constant, since Au is the volume flowing through the section in unit time.

If a small element of the tube is taken, of length dx, so that the pressure difference between its ends is dP, then the volume leaving the element in unit time is, from equation (7.4)

$$V = -\frac{\pi a^4}{8\eta} \cdot \frac{dP}{dx} \qquad \qquad \ldots\ldots (7.31)$$

The negative sign indicates that P decreases as x increases. If the volume of gas entering the tube in unit time is V_1 at a pressure P_1, then

$$P_1 V_1 = PV$$

$$P_1 V_1 = -\frac{\pi a^4}{8\eta} \cdot P \cdot \frac{dP}{dx} \qquad \qquad \ldots\ldots (7.32)$$

$$\int_0^l P_1 V_1 dx = -\frac{\pi a^4}{8\eta} \int_{P_1}^{P_2} P\, dP$$

where P_2 is the pressure at the outlet end of the tube of length l.

Hence $\qquad\qquad P_1 V_1 = (P_1^2 - P_2^2)\dfrac{\pi a^4}{16\eta l} \qquad\qquad \ldots\ldots (7.33)$

In the above formula no account has been taken of the slipping of the gas in contact with the walls of the tube. The gas in contact with the tube does not remain at rest but slips, so that the gas flows as if the radius of the tube were increased by a factor λ, where λ is the mean free path. Thus the radius of the tube must be modified to $(a + \lambda)$, where λ is very small. Expanding $(a + \lambda)^4$ by the binomial theorem and neglecting higher powers, Poiseuille's modified formula becomes

$$P_1 V_1 = (P_1^2 - P_2^2)\frac{\pi a^4}{16\eta l}\left(1 + \frac{4\lambda}{a}\right) \qquad \ldots (7.34)$$

7.11 DETERMINATION OF VISCOSITY OF GASES
7.11.1 Viscosity of air

The simplest method available for the determination of the viscosity of air is probably that which utilizes a U-tube, one arm of which is a capillary tube of length l and radius a. In the other arm of the U-tube is a mercury pellet, of weight mg. The mercury pellet is adjusted to be level with a fixed mark on the arm of the tube and it is then allowed to fall vertically from this mark. The time taken for the pellet to fall to a second fixed mark lower down on the arm of the tube, is noted. If the volume between the two fixed marks on the wide tube is Q, then, if the time of fall of the pellet is t, the volume of air passing through the capillary tube per second is Q/t. The pressure at the inlet end of the capillary tube is $P_0 + mg/A - \delta$, where P_0 is atmospheric pressure, mg/A is the pressure due to the mercury pellet, A being the cross-sectional area of the wide tube, and δ represents a reduction in the excess pressure due to friction between the walls of the tube and the mercury pellet. It is often called the sticking coefficient. Inserting these values of P_1 and V_1 in equation (7.33)

$$\left(P_0 + \frac{mg}{A} - \delta\right)\frac{Q}{t} = \left[\left(P_0 + \frac{mg}{A} - \delta\right)^2 - P_0^2\right]\frac{\pi a^4}{16\eta l} \qquad \ldots (7.35)$$

i.e. $\quad \dfrac{16\eta l Q}{\pi a^4 t} = \left[\left(P_0 + \dfrac{mg}{A} - \delta\right) - P_0\left(1 + \dfrac{\frac{mg}{A} - \delta}{P_0}\right)^{-1}\right]$

Expanding by the binomial and neglecting terms in δ/P_0

$$\frac{16\eta l Q}{\pi a^4 t} = \left[P_0 + \frac{mg}{A} - \delta - P_0 + \frac{mg}{A} - \delta - \frac{\left(\frac{mg}{A}\right)^2}{P_0}\right]$$

$$\frac{16\eta l Q}{\pi a^4 t} = \left[\frac{2mg}{A} - \frac{\left(\frac{mg}{A}\right)^2}{P_0} - 2\delta\right]$$

208

Hence

$$\frac{8\eta l Q}{\pi a^4} = \left[\frac{mg}{A} - \frac{\left(\frac{mg}{A}\right)^2}{2P_0} - \delta\right] t = \text{constant} \qquad \ldots (7.36)$$

The sticking coefficient can be eliminated by carrying out the experiment using mercury pellets of different weights. If the times of fall are t and t_1 for pellets of weight mg and $m_1 g$ respectively, then, from equation (7.36)

$$\left[\frac{mg}{A} - \frac{\left(\frac{mg}{A}\right)^2}{2P_0} - \delta\right] t = \left[\frac{m_1 g}{A} - \frac{\left(\frac{m_1 g}{A}\right)^2}{2P_0} - \delta\right] t_1 \quad \ldots (7.37)$$

and thus δ may be determined in terms of known quantities.

7.11.2 Constant volume method

A number of experimental determinations of the viscosity of gases have been made at constant volume. These are all basically similar and the one due to Edwards is described below. The apparatus is shown in *Figure 7.10*. The gas is contained in a large bulb, volume Q, which is, in turn, contained in a constant temperature enclosure. The gas is at a pressure P_1, greater than atmospheric P_0, and this is determined by the mercury manometer. A capillary tube, of length l and radius a, leads from the bulb to the atmosphere *via* the tap T_2. With the tap T_1 closed, the tap T_2 is opened so that the gas flows through the capillary tube for t s. T_2 is closed and T_1 opened so that the new gas pressure P_2 in the bulb can be measured. If the volume of gas entering the capillary tube per second at time t is V_1 when the pressure in the apparatus is P, then, for a slow rate of flow

$$PQ = (P + dP)(Q + V_1\, dt) \qquad \ldots (7.38)$$

where, in the interval dt, the pressure changes to $P + dP$. From equation (7.38)

$$PV_1 \simeq -Q\frac{dP}{dt}$$

However, from equation (7.33)

$$PV_1 = (P^2 - P_0^2)\frac{\pi a^4}{16\eta l} \qquad \ldots (7.39)$$

209

Hence $\dfrac{\pi a^4}{16\eta l Q} \displaystyle\int_0^t dt \qquad = -\displaystyle\int_{P_1}^{P_2} \dfrac{dP}{P^2 - P_0^2}$

$$= -\dfrac{1}{2P_0}\left[\int_{P_1}^{P_2} \dfrac{dP}{P - P_0} - \int_{P_1}^{P_2} \dfrac{dP}{P + P_0}\right]$$

$$= -\dfrac{1}{2P_0}\log \dfrac{(P_2 - P_0)}{(P_1 - P_0)}\cdot\dfrac{(P_1 + P_0)}{(P_2 + P_0)}$$

therefore $\qquad \dfrac{\pi a^4 t}{16\eta l Q} = +\dfrac{1}{2P_0}\log \dfrac{(P_1 - P_0)(P_2 + P_0)}{(P_2 - P_0)(P_1 + P_0)}$

$$\dots\dots (7.40)$$

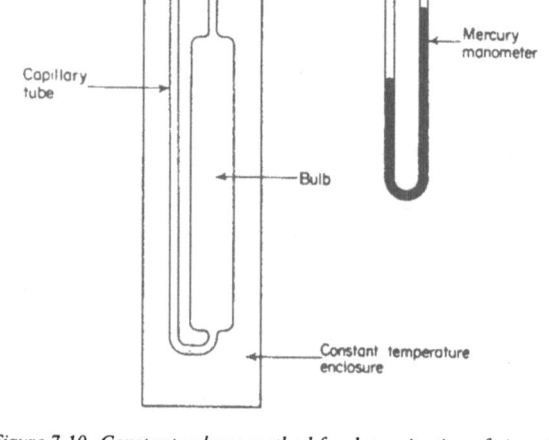

Figure 7.10. *Constant volume method for determination of viscosity*

If the barometric height is h_0 and the difference in the heights of the mercury manometer, at the beginning and end of the experiment, are h_1 and h_2 respectively, then $P_0 = h_0\rho g$, $P_1 = (h_0 + h_1)\rho g$ and $P_2 = (h_0 + h_2)\rho g$, where ρ is the density of mercury. Substituting in equation (7.40)

$$\dfrac{h_0 \pi a^4 t}{8\eta l Q} = \log\left(\dfrac{h_1}{h_2}\cdot\dfrac{2h_0 + h_2}{2h_0 + h_1}\right) \qquad \dots\dots (7.41)$$

This equation can be modified to take into account the slipping of the gas, as in equation (7.34). From equation (7.41) it is a simple matter to calculate η since all the other quantities are experimental constants or can be measured.

7.11.3 Viscosity of hydrogen or oxygen at room temperature

This experiment was suggested by Lehfeldt and his apparatus is shown in *Figure 7.11*. It consists of a water voltameter through which a suitable steady current I is passed by means of the circuit shown. Oxygen or hydrogen is led off from the voltameter, through the drying tube containing calcium chloride and enters a capillary tube of length l and radius a. The pressure of gas entering the capillary tube, P_1, is determined by the light oil manometer. The initial pressure of gas in the apparatus will build up until a state of equilibrium is reached when the volume of gas entering the capillary tube per second is just equal to the volume leaving the capillary tube per second at atmospheric pressure P_0. If the electrochemical equivalent of oxygen, in grammes per coulomb, is Z, then the mass of gas liberated per second is ZI g. At s.t.p. this occupies a volume of $ZI . 22\cdot4 . 10^3/32$ cm^3.

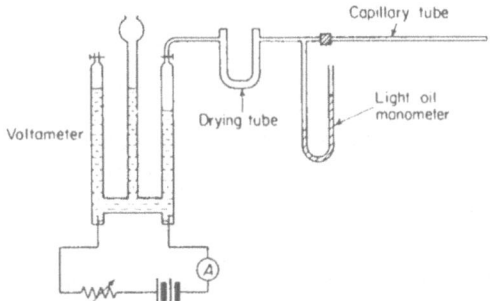

Figure 7.11. Viscosity of oxygen or hydrogen at room temperature

If the experiment is carried out at a temperature of $t°$C and the atmospheric pressure is P_0, then the gas liberated will occupy a volume, V_2, given by

$$ZI \times \frac{22\cdot4 \times 10^3}{32} \times \frac{t + 273}{273} \times \frac{760}{P_0} \qquad \dots\, (7.42)$$

Now, from equation (7.33)

211

$$P_1 V_1 = \frac{\pi a^4}{8\eta l} \cdot \frac{P_1^2 - P_2^2}{2} = P_2 V_2 \qquad \dots (7.43)$$

Here V_2 is the volume of gas escaping per second at the pressure P_0 when equilibrium is reached and thus it is also the volume of gas generated per second, measured at the pressure P_0. Hence, if the difference in the heights of the manometer arms is h and the density of the oil is ρ, since $P_1 = P_0 + \rho g h$ and $P_2 = P_0$, from equations (7.42) and (7.43)

$$V_2 = ZI \cdot \frac{22 \cdot 4 \times 10^3}{32} \cdot \frac{t + 273}{273} \cdot \frac{760}{P_0} = \frac{\pi a^4}{8\eta l} \frac{(2P_0 + \rho g h) \rho g h}{2P_0}$$

$$\dots (7.44)$$

Hence

$$\eta = \frac{\pi a^4}{8lZI} \cdot \frac{(2P_0 + \rho g h) \rho g h}{2P_0} \cdot \frac{32}{22 \cdot 4 \times 10^3} \cdot \frac{273}{t + 273} \cdot \frac{P_0}{760} \qquad \dots (7.45)$$

The experiment is carried out by using values of I over a suitable range and measuring the corresponding values of h. To obtain accurate results with this equipment it is important to make quite sure that the apparatus is leak-proof.

7.11.4 Rankine's method for the determination of the viscosity of a gas

The experiment designed by Rankine was used initially for the determination of the viscosity of the rare gases. It is very suitable for the determination of the viscosity of any gas, particularly when it is available only in small quantities.

The apparatus is shown in *Figure 7.12*. It consists of a closed tube, *ABCDEF*, a few millimetres in diameter, joined to a capillary tube. Two marks, *C* and *D* are fixed so that the volume of *ABC* is equal to that of *FED*, V'. A mercury pellet, of weight mg, is used to produce excess pressure and the time, t, for it to fall from *C* to *D* is noted.

The theory of the experiment is as follows. Let the total volume of the apparatus, unoccupied by the mercury pellet, be V. When the apparatus is initially horizontal, let the uniform pressure throughout be P and let the density of the gas at this pressure be $P\rho$, where ρ is the density of the gas at unit pressure. Thus the mass of gas enclosed, which is always the same, is $PV\rho$. The apparatus is placed in the vertical position to cause the pellet to fall from *C* to *D*. Initially the pressure in *ABC* is p_1 while that in *CEF* is $P_1 = p_1 + mg/A$, where A is the cross-sectional area of the tube. Since the mass of gas remains constant

$$PV\rho = p_1 V'\rho + \left(p_1 + \frac{mg}{A}\right)(V - V')\rho$$

from which

$$p_1 = P - \frac{mg}{A} + \frac{mg}{A}\frac{V'}{V} \qquad \qquad \ldots(7.46)$$

and

$$p_1 = P + \frac{mg}{A} \cdot \frac{V'}{V} \qquad \qquad \ldots(7.47)$$

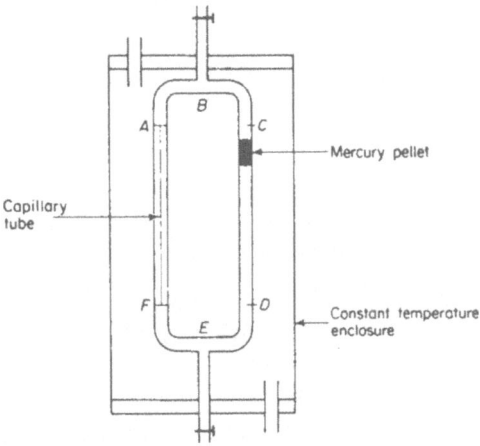

Figure 7.12. Rankine's apparatus

When the mercury pellet has fallen to D, the pressures in ACD become,

$$p_2 = P - \frac{mg}{A} + \frac{mg}{A} - \frac{V - V'}{V}$$

$$= P - \frac{mg}{A} \cdot \frac{V'}{V} \qquad \qquad \ldots(7.48)$$

and in DEF

$$p_2 = P + \frac{mg}{A} \cdot \frac{V - V'}{V} \qquad \qquad \ldots(7.49)$$

213

Now the mass of gas which has passed through the capillary is

$$P_1(V - V')\rho - P_2V'\rho = \left(P + \frac{mg}{A} \cdot \frac{V'}{V}\right)(V - V')\rho$$

$$- \left(P + \frac{mg}{A} \cdot \frac{V - V'}{V}\right)V'\rho = P(V - 2V')\rho$$

Thus the average rate of flow of gas through the capillary in grammes per second is

$$P\frac{(V - 2V')}{t}\rho \qquad \qquad \dots (7.50)$$

Now, applying equation (7.33), the mass per second of gas initially entering the capillary tube is

$$P_1V_1\rho = \frac{(P_1^2 - p_1^2)\,\pi a^4}{16\eta l}\rho = \frac{\pi a^4\rho}{16\eta l} \cdot \frac{mg}{A}\left(2P - \frac{mg}{A} + \frac{2mg}{A} \cdot \frac{V'}{V}\right)$$

$$\dots (7.51)$$

where V_1 is the initial volume per second.

The mass per second of gas finally entering the capillary tube, where V_2 is the final volume per second, is

$$P_2V_2\rho = \frac{(P_2^2 - p_2^2)\,\pi a^4}{16\eta l}\rho = \frac{\pi a^4\rho}{16\eta l} \cdot \frac{mg}{A}\left(2P + \frac{mg}{A} - \frac{2mg}{A} \cdot \frac{V'}{V}\right)$$

$$\dots (7.52)$$

Thus, from equations (7.51) and (7.52), the average rate of flow of gas through the capillary, in grammes per second, is

$$\frac{\pi a^4\rho}{16\eta l} \cdot \frac{mg}{A} \cdot 2P \qquad \qquad \dots (7.53)$$

But this is equal to equation (7.50), so that

$$\frac{P(V - 2V')}{t}\rho = \frac{\pi a^4\rho}{16\eta l} \cdot \frac{mg}{A} \cdot 2P$$

Hence $\qquad \dfrac{(V - 2V')}{t} = \dfrac{\pi a^4 mg}{8\eta lA} \qquad \qquad \dots (7.54)$

where $V - 2V'$ is the volume of the tube between the two fixed marks C and D, and A is the area of cross-section of the tube. A correction can be made for the friction between the walls of the

214

tube and the mercury pellet as described in Subsection 7.11.1. This experiment is very suitable for measuring the variation of viscosity with both temperature and pressure and, in fact, using this apparatus, Rankine showed that the viscosity of a gas is indeed independent of its pressure, as predicted by kinetic theory.

7.11.5 Millikan's method for absolute determination of the viscosity of a gas

When Millikan was carrying out his famous experiments to determine the electronic charge he found that the results of his experiments were limited in accuracy by the accuracy of the available value for the viscosity of air. Accordingly he decided to measure the viscosity of air as accurately as possible and to do this he used a form of the rotation viscometer discussed in Subsection 7.7.1. This technique was chosen in preference to other methods since errors could be more readily eliminated. In Subsection 7.7.1 it was shown that the torque exerted on the inner cylinder, equation (7.13), is

$$\tau_1 = \frac{4\pi\eta l\Omega a^2 b^2}{(b^2 - a^2)}$$

If the inner cylinder is suspended from a torsion head, then it will experience a steady angular deflection of θ from the equilibrium position, due to the action of the torque. If the couple per unit twist in the torsion wire is c, then

$$\tau_1 = c\theta \qquad \qquad \dots (7.55)$$

If the moment of inertia of the inner cylinder about its axis of suspension is I and T is the period of oscillation, then $T = 2\pi\sqrt{(I/c)}$, and hence

$$\tau_1 = \frac{4\pi^2 I\theta}{T^2}$$

so that

$$\eta = \frac{\pi I(b^2 - a^2)\,\theta}{T^2 la^2 b^2 \Omega} \qquad \qquad \dots (7.56)$$

Millikan's apparatus is shown in *Figure 7.13*. The inner cylinder was made of very thin brass in order to reduce its mass. It was suspended from a torsion head as shown. Its period of oscillation was found and also its moment of inertia. The outer cylinder was rotated at a constant rate and the steady angular deflection of the

215

inner cylinder measured. End effects were reduced to negligible proportions by the coaxial guard rings as shown.

A similar experiment was carried out in 1939 by Bearden who actually rotated the inner cylinder. In his apparatus provision was made for filling the space between the cylinders with any gas at a required pressure. This work has provided accurate information on the viscosity of gases.

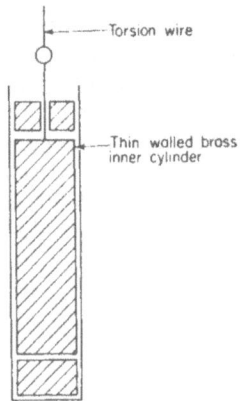

Figure 7.13. Millikan's apparatus

7.11.6 Maxwell's oscillating disk viscometer

In the oscillation viscometer a suspended body oscillates about a vertical axis of symmetry and the rate of damping is approximately proportional to the viscosity of the gas surrounding the body. In the oscillation viscometer there is no expansion of the gas as in viscometers based on the flow of the gas. Maxwell used the oscillating disk viscometer to investigate the variation of the viscosity of a gas with pressure and found that viscosity is independent of pressure as he had predicted from kinetic theory.

A modification of the oscillating disk viscometer, due to Vogel, has provided a relatively simple method of determining the viscosity of gases at low temperatures. His apparatus is shown diagrammatically in *Figure 7.14*.

The apparatus consists of a thin glass disk attached by a nickel wire to a mirror. The whole is then suspended by a platinum wire from the hook attached to the torsion head. Two fixed glass plates are held rigidly either side of the glass disk by means of the springs and separating blocks. The suspension system is set in motion by means of the astatic pair of magnets. The whole suspension system is contained in a glass vessel provided with a viewing window, so that the oscillations may be observed. The vessel is evacuated and the gas being examined is allowed to flow in through the inlet. The lower part of the glass vessel is placed in a cryostat at the required temperature. The long neck of the apparatus ensures that the platinum suspension wire is sufficiently far removed from the cold part of the apparatus for its elastic properties to remain constant.

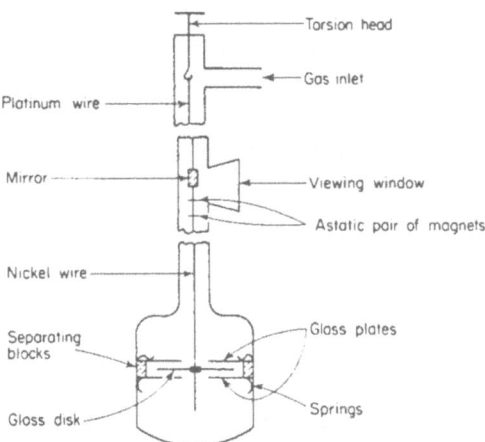

Figure 7.14. Oscillating disk viscometer

The theory of the oscillating disk is somewhat complicated and also several uncertain corrections are necessary. This means that the instrument is not really suitable for absolute measurements and it is normally used for comparison. In his experiments, Vogel assumed the value of the viscosity of air at 0°C which had been determined by other methods. The use of this apparatus has been extended to measurements on gases at high temperatures and it has also been used for viscosity measurements on molten metals.

7.12 EFFECT OF TEMPERATURE AND PRESSURE ON THE VISCOSITY OF GASES

The viscosity of gases increases as the temperature increases. On the basis of elementary kinetic theory the relation

$$\eta = kT^{\frac{1}{2}}$$

can be derived, where T is the absolute temperature and k is a

217

constant. A more accurate formula was derived by Sutherland who took into account the attractive force between the neighbouring gas molecules. This takes account of the fact that molecular collisions in a real gas are more abundant than in an ideal gas at the same temperature. Sutherland's formula relating temperature and viscosity is

$$\frac{\eta_t}{\eta_0} = \frac{273 + C}{T + C} \left(\frac{T}{273} \right)^{\frac{3}{2}} \qquad \ldots (7.57)$$

where η_t and η_0 are the viscosities at $T°$ absolute and 0°C respectively and C is Sutherland's constant for a particular gas. This formula agrees well with experimental results obtained over a wide range of temperature but it fails at both very high and very low temperatures.

According to the kinetic theory the viscosity of gases should be independent of pressure and Maxwell, and later, other workers, showed that this prediction was true over a wide range of pressures. At low pressures, however, where the mean free path of the molecules becomes of the same order of magnitude as the dimensions of the containing vessel, kinetic theory shows that viscosity is directly proportional to the pressure and again, this theoretical prediction has been confirmed by experimental work. At high pressures it is found experimentally that the viscosity increases with an increase in pressure and it has been shown that this is due to the Hagenbach correction (Section 7.3) becoming increasingly important.

7.13 NON-NEWTONIAN LIQUIDS

In Section 7.1 a liquid was defined as a material which is incapable of sustaining a shearing stress. Newton's law of viscous flow, equation (7.1), was formulated on the assumption that for a fluid moving with streamline motion the shearing stress at any point is directly proportional to the rate of shear, measured by the velocity gradient. Liquids exhibiting such proportionality are normally described as Newtonian liquids. If a liquid does not exhibit this proportionality then it is called a non-Newtonian liquid.

Nearly all pure liquids are Newtonian but it is found that many impure liquids, such as colloids, solid-liquid mixtures, solutions of long chain molecules, etc., are non-Newtonian. In these impure, heterogeneous liquids, the shearing stress is not proportional to the rate of shear and the shearing stress increases either more or less rapidly than the velocity gradient. The study of such liquids and flow deformations generally is given the name 'rheology' and considerable work has been carried out in this field.

Non-Newtonian liquids in which particles are floating in the liquid, are called sols. Alternatively, non-Newtonian liquids may be in the form of a gel, where the liquid has some internal structure rendering it semi-solid. If the concentration of material in suspension in the liquid is high, it is found that the liquid can withstand small shearing stresses without moving, in contrast to Newtonian liquids which move under the smallest applied shearing stress. An increase in the shearing stress will ultimately cause the liquid to flow and the stress at the point where flow begins is known as the yield value for the liquid. This behaviour is exhibited by stiff pastes and jellies.

Many substances, such as paints, creams, honey, etc., which possess gel characteristics, are found temporarily to become sols if they are agitated. If allowed to stand they ultimately revert to their former consistency. This effect, known as thixotropy, is due to a disorientation of the molecules from their orderly arrangement in the original gel by the agitation. When the agitation is stopped the molecules gradually resume their orderly structure.

Thixotropy is a property which should be possessed by paints so that they can be brushed freely over a surface when stirred but will settle to a uniform texture when left.

7.13.1 Bingham plastic flow

Much of the early work on non-Newtonian liquids and the plastic state generally, was performed by Bingham and he defined an ideal plastic solid as a material for which the rate of flow was zero for shearing stresses below a critical value and proportional to

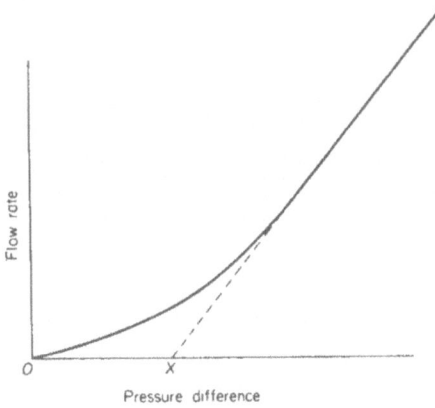

Figure 7.15. Bingham plastic flow

219

the excess stress above this value. Bingham introduced this term in order to describe the rheological behaviour of concentrated clay suspensions. *Figure 7.15* illustrates the type of curve often obtained when a non-Newtonian liquid is forced through a capillary tube under a pressure difference between its ends. It can be seen that the liquid does not suddenly begin to flow at a critical value of the stress as in true Bingham plastic flow. An approximation to Bingham plastic flow can be made, however, by making the assumption that there is no yield for a stress S_0 below the yield point X, and that for an excess applied stress, $S - S_0$, above S_0, the flow is subsequently Newtonian. On this basis an equation can be derived relating the pressure difference and the volume of liquid flowing through the tube. If the pressure difference is P, the critical stress, S_0, occurs at a distance r_0 from the centre of the tube, where

$$P\pi r_0^2 = 2\pi r_0 l S_0 \qquad \dots (7.58)$$

l being the length of the capillary tube.

Hence the critical stress S_0 is equal to $Pr_0/2l$. At a distance $r > r_0$ from the centre of the tube the shear stress S is equal to $Pr/2l$ and the net stress which does work against viscous forces is $S - S_0$. If the velocity of the liquid at the radial distance r is u, then, assuming Newtonian flow

$$\eta \frac{du}{dr} = - \frac{P}{2l}(r - r_0) \qquad \dots (7.59)$$

where η is the coefficient of viscosity for the liquid in the region where the rate of flow is proportional to the excess stress. Integration of equation (7.59) gives

$$u = \frac{P}{2\eta l}\left[\frac{a^2 - r^2}{2} - r_0(a - r)\right]$$

where a is the radius of the capillary tube. When $r = r_0$

$$u = u_0 = \frac{P}{2\eta l}\left[\frac{a^2 - r_0^2}{2} - r_0(a + r_0)\right]$$

$$u_0 = \frac{P}{4\eta l}(a - r_0)^2$$

When r is less than r_0 the velocity is constant at u_0 so that effectively a solid cylindrical core of liquid flows through the tube with velocity u_0.

If the volume of liquid flowing through the remainder of the tube in unit time is V_1, then

$$dV_1 = 2\pi ru\,dr = \frac{P\pi}{\eta l}\left(\frac{a^2 - r^2}{2} - r_0 a + r_0 r\right) r\,dr$$

and

$$V_1 = \frac{P\pi}{8\eta l}(a^4 - \tfrac{4}{3}a^3 r_0 - 2a^2 r_0^2 + 4ar_0^3 - \tfrac{5}{3}r_0^4)$$

The volume (V_0) flowing in unit time through the central part of the tube, is given by

$$V_0 = \pi r_0^2 u_0$$

$$= \frac{P\pi}{8\eta l}[2a^2 r_0^2 - 4ar_0^3 + 2r_0^4]$$

Hence, the total volume of liquid flowing through the tube in unit time is

$$V = V_0 + V_1 = \frac{P\pi}{8\eta l}\left[a^4 - \frac{4}{3}a^3 r_0 + \frac{r_0^4}{3}\right] \qquad \ldots (7.60)$$

Equation (7.60) is known as the Buckingham–Reiner equation and is applicable to a wide range of non-Newtonian liquids such as clay and soil pastes.

7.13.2 Measurement of rheological properties

A study of the viscous properties of non-Newtonian liquids involves subjecting the liquid to varying rates of shear and measuring the effect produced for each particular value of the shear rate. In ordinary viscometers, for Newtonian liquids, the rate of shear varies between certain limits, and hence the measured flow rate is produced by many different shear rates. It is impossible to give a value for the apparent viscosity coefficient for the liquid at a particular rate of shear and any value given represents an average for the different shear rates.

While it is extremely difficult to deduce the exact variation of the apparent coefficient of viscosity of the non-Newtonian liquid with rate of shear, it is possible to examine, qualitatively, how the apparent viscosity varies with the mean rate of shear and to compare the behaviour of the liquid with a true Newtonian liquid.

The Couette or concentric cylinder viscometer (*illustrated in Figure 7.16*) is often used in an examination of the behaviour of non-Newtonian liquids. The liquid is contained in the gap between the cylinders. The outer cylinder is rotated and the rotational speed is a measure of the rate of shear. The inner cylinder is held in a fixed position and the torque necessary to maintain its fixed position is, after making certain corrections, a measure of the shear stress.

If the ratio of the radii of the two cylinders is nearly unity the shear stress and shear rate are nearly uniform throughout the liquid and a shear stress versus shear rate plot can be made direct

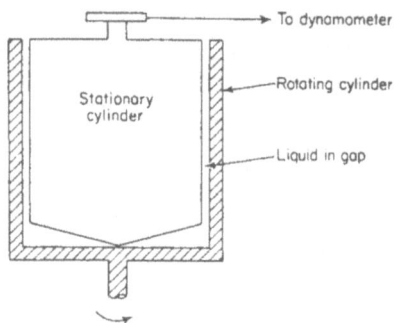

Figure 7.16. Couette viscometer

from the experimental results. Due to viscous heating, a change in the rheological properties of the liquid may occur, and this effect must be taken into account.

Another instrument used for the measurement of apparent viscosities is the cone and plate device, shown in *Figure 7.17*. The

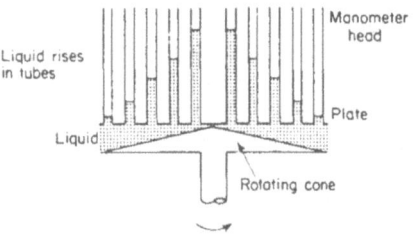

Figure 7.17. Cone and plate device

liquid is sheared between the rotating cone and the fixed plate. The shear rate–shear stress relationship is obtained from measurements of the rotational speed and torque required to drive the cone. If the angle between the cone and plate is small the shear stress is nearly uniform throughout the liquid. If the instrument is used in the study of normal stresses, the manometer head is used and the instrument is called a rheogoniometer. A correction for the temperature rise due to viscous heating is necessary as with the Couette instrument.

A capillary tube instrument has also been used in the measurement of the properties of non-Newtonian liquids. The choice of instrument for the study of the behaviour of non-Newtonian liquids is determined by the consistency of the liquid and by the range over which the shear rate is to be varied.

Some non-Newtonian liquids, such as aluminium dilaurate in a hydrocarbon oil, possess remarkable properties, and a great deal of work is currently being done both theoretically and experimentally, in order to describe quantitatively their behaviour.

8

HYDRODYNAMICS

8.1 INTRODUCTION

CONSIDER the motion of any entity such as matter, heat, electricity, etc. Assume the entity to be continuous and assume also that the properties of the smallest portion of it are the same as those of the entity in bulk. Let u, v and w be the components, parallel to the coordinate axes, of the velocity vector v at the point (x, y, z) at the time t. Assume u, v and w to be finite and continuous functions of x, y and z and $\partial u/\partial x$, $\partial v/\partial x$, $\partial w/\partial x$, also to be finite. Let the motion in space be opposed by obstacles so that, on considering an area perpendicular to the direction of motion, the space available for the passage of the entity is less than if there were no obstacles and let K, the permeability, be a measure of the obstruction, i.e. K is the ratio of the areas available with and without the obstruction.

The increase in mass of an entity within any closed boundary is equal to the excess which flows in over that flowing out together with any entity created in the volume. Let m be the mass created in unit volume in unit time, K_x, K_y and K_z the permeabilities and Q_x, Q_y, Q_z, the entity densities parallel to the coordinate axes, Q being the average density at the point (x, y, z). With this point as centre construct a small parallelepiped of edges δx, δy, δz, as in *Figure 8.1*.

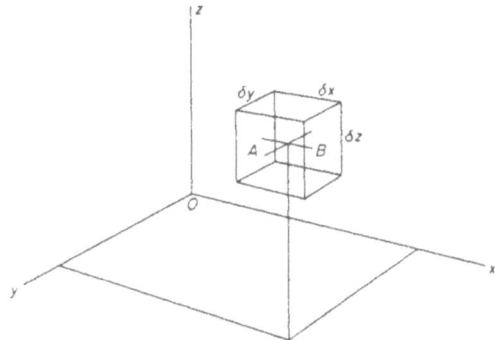

Figure 8.1. Equation of continuity in cartesian coordinates

224

Now the entity flowing in through the face A in time δt is

$$\left(K_x - \frac{\partial K_x}{\partial x}.\frac{\delta x}{2}\right)\left(Q_x - \frac{\partial Q_x}{\partial x}.\frac{\delta x}{2}\right)\left(u - \frac{\partial u}{\partial x}.\frac{\delta x}{2}\right)\delta y.\delta z.\delta t$$

$$\ldots (8.1)$$

The amount flowing out through the face B is

$$\left(K_x + \frac{\partial K_x}{\partial x}.\frac{\delta x}{2}\right)\left(Q_x + \frac{\partial Q_x}{\partial x}.\frac{\delta x}{2}\right)\left(u + \frac{\partial y}{\partial x}.\frac{\delta x}{2}\right)\delta y.\delta z.\delta t$$

$$\ldots (8.2)$$

Thus the gain in volume due to flow along the x axis is approximately

$$-\frac{\partial}{\partial x}(K_x.Q_x.u)\,\delta x\delta y\delta z\delta t \qquad \ldots (8.3)$$

Hence the total gain in entity is

$$\left[m - \left\{\frac{\partial}{\partial x}(K_x.Q_x.u) + \frac{\partial}{\partial y}(K_y.Q_y.v) + \frac{\partial}{\partial z}(K_z.Q_z.w)\right\}\right]$$

$$\delta x.\delta y.\delta z.\delta t \qquad \ldots (8.4)$$

The original mass within the volume is $Q\delta x\delta y\delta z$ so that the gain in time δt is $(\partial Q/\partial t).\delta x\delta y\delta z\delta t$ therefore

$$\frac{\partial Q}{\partial t} - m + \frac{\partial}{\partial x}(K_x.Q_x.u) + \frac{\partial}{\partial y}(K_y.Q_y.v) + \frac{\partial}{\partial z}(K_z.Q_z.w) = 0$$

$$\ldots (8.5)$$

Equation (8.5) is the general equation of continuity. When applied to matter, since matter can neither be created nor destroyed, $m = 0$. Also, $K_x = K_y = K_z = 1$ and $Q_x = Q_y = Q_z = \rho$, the density of matter. Thus equation (8.5) becomes

$$\frac{\partial \rho}{\partial t} + \frac{\partial}{\partial x}(\rho u) + \frac{\partial}{\partial y}(\rho v) + \frac{\partial}{\partial z}(\rho w) = 0 \qquad \ldots (8.6)$$

This is the hydrodynamical equation of continuity. It is written in vector terms as

$$\frac{\partial \rho}{\partial t} + \text{div}\,(\rho \boldsymbol{v}) = 0 \qquad \ldots (8.7)$$

8.2 EULER'S EQUATION OF MOTION

Consider the case where the entity refers to the momentum of incompressible matter. Consider forces acting along the x axis

225

which cause the matter to have a velocity u in the x direction. Under such circumstances, $Q = \rho u$ and the equation of continuity becomes

$$\frac{\partial}{\partial t}(\rho u) + \frac{\partial}{\partial x}(\rho u^2) + \frac{\partial}{\partial y}(\rho uv) + \frac{\partial}{\partial z}(\rho uw) - m = 0 \quad \ldots (8.8)$$

For matter, ρ is independent of direction and thus, from equations (8.6) and (8.8)

$$\frac{\partial u}{\partial t} + u\frac{\partial u}{\partial x} + v\frac{\partial u}{\partial y} + w\frac{\partial u}{\partial z} - \frac{m}{\rho} = 0 \quad \ldots (8.9)$$

Now the momentum may be produced by either of the following:
  (1) An impressed force at a distance acting on the mass.
  (2) Pressure acting on the boundary surface.

Let the components of the impressed force per unit mass at the point (x, y, z) be F_x, F_y, F_z and let p be the pressure at this point. The total force acting on the volume $\delta x \delta y \delta z$ in the x direction is then

$$-\frac{\partial p}{\partial x}.\delta x \delta y \delta z + F_x \rho \delta x \delta y \delta z$$

Therefore the momentum created per unit volume per second is

$$m = F_x \rho - \frac{\partial p}{\partial x} \quad \ldots (8.10)$$

Substituting this value of m in equation (8.9)

$$\frac{\partial u}{\partial t} + u\frac{\partial u}{\partial x} + v\frac{\partial u}{\partial y} + w\frac{\partial u}{\partial z} - F_x + \frac{1}{\rho}.\frac{\partial p}{\partial x} = 0 \quad \ldots (8.11)$$

Similarly

$$\frac{\partial v}{\partial t} + u\frac{\partial v}{\partial x} + v\frac{\partial v}{\partial y} + w\frac{\partial v}{\partial z} - F_y + \frac{1}{\rho}.\frac{\partial p}{\partial y} = 0$$

and

$$\frac{\partial w}{\partial t} + u\frac{\partial w}{\partial x} + v\frac{\partial w}{\partial y} + w\frac{\partial w}{\partial z} - F_z + \frac{1}{\rho}.\frac{\partial p}{\partial y} = 0$$

These equations are known as Euler's equations of motion. In vector terms the equations are more simply expressed as

$$\frac{\partial \mathbf{v}}{\partial t} + (\mathbf{v}\,\text{grad})\,\mathbf{v} = -\frac{1}{\rho}\,\text{grad}\,p + \mathbf{F} \quad \ldots (8.12)$$

226

If the impressed force F is due to gravitation then $F = g$, the vector acceleration of gravity.

Equation (8.12) only applies to ideal fluids since dissipative effects have not been considered. It forms the basis of many calculations of fluid flow.

8.3 BERNOULLI'S THEOREM

Frequently the component velocities may be expressed in terms of a single function ϕ known as the velocity potential. Thus, $u, v, w = -\partial\phi/\partial x, -\partial\phi/\partial y, -\partial\phi/\partial z$, respectively.

Euler's equations of motion may then be written as

$$-\frac{\partial^2\phi}{\partial x.\partial t} + u\frac{\partial y}{\partial x} + v\frac{\partial v}{\partial x} + w\frac{\partial w}{\partial x} = -\frac{\partial\Omega}{\partial x} - \frac{1}{\rho}\cdot\frac{\partial p}{\partial x}\dots\text{etc.,}\dots\text{(8.13)}$$

since

$$\frac{\partial v}{\partial z} = \frac{\partial w}{\partial y}, \quad \frac{\partial w}{\partial x} = \frac{\partial u}{\partial z}, \quad \frac{\partial u}{\partial y} = \frac{\partial v}{\partial x}$$

and $F = -\partial\Omega/\partial x$, Ω denoting the potential energy per unit mass at a point (x, y, z) in respect of the forces acting at a distance. Integration of equation (8.13) gives

$$\int\frac{dp}{\rho} = \frac{\partial\phi}{\partial t} - \Omega - q^2/2 + F(t) \qquad \dots\text{(8.14)}$$

where $q = (u^2 + v^2 + w^2)^{\frac{1}{2}}$ and $F(t)$ is an arbitrary function of time. Under steady conditions the velocity of a fluid is the same in direction and magnitude at all times, i.e.

$$\frac{\partial v}{\partial t} = 0 \qquad \text{or} \qquad \frac{\partial u}{\partial t} = \frac{\partial v}{\partial t} = \frac{\partial w}{\partial t} = 0$$

Hence equation (8.14) becomes, under steady conditions

$$\int\frac{dp}{\rho} = -\Omega - \frac{q^2}{2} + \text{constant} \qquad \dots\text{(8.15)}$$

Now for steady motion the streamlines coincide with the paths of the fluid particles, a streamline being defined as a line such that the tangent to the line at any point is in the direction of the fluid velocity at that point. Thus equation (8.15) is a law of variation of pressure along a streamline and it is known as Bernoulli's theorem. The integration is to be made along the streamline corresponding to the constant in the equation.

HYDRODYNAMICS

Bernoulli's theorem may also be deduced from the principle of the conservation of energy. Consider a tube of flow in the liquid, the boundary surface of which is formed by streamlines. Assume the fluid to be incompressible and ignore viscous forces. Let p_1 be the pressure, q_1 the velocity, and Ω_1 the potential due to external forces at the point A where the cross-sectional area is α_1 (*Figure 8.2*).

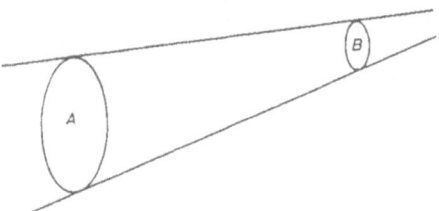

Figure 8.2. Derivation of Bernoulli's theorem

Let the same quantities at another point B be p_2, q_2, Ω_2 and α_2 respectively. Since the mass of fluid contained between the normal sections of a tube is constant, the same mass passes every normal section in unit time, i.e.

$$\alpha_1 q_1 = \alpha_2 q_2$$

The work done on the mass entering A per unit time is $p_1\alpha_1 q_1$ and on the mass leaving at B, $p_2\alpha_2 q_2$. The mass entering at A per unit time has kinetic energy $\frac{1}{2}\alpha_1 q_1 \rho q_1^2$ and that leaving at B, $\frac{1}{2}\alpha_2 q_2 \rho q_2^2$.

The mass at A has potential energy $\Omega_1\alpha_1 q_1\rho$ and that at B, $\Omega_2\alpha_2 q_2\rho$. Since the motion is steady, by the conservation of energy, the energy within the tube remains constant. Thus

$$\alpha_1 q_1\rho\left(\Omega_1 + \frac{q_1^2}{2}\right) + p_1\alpha_1 q_1 = \alpha_2 q_2\rho\left(\Omega_2 + \frac{q_2^2}{2}\right) + p_2\alpha_2 q_2 \ldots (8.16)$$

i.e.
$$\frac{p_1}{\rho} + \frac{q_1^2}{2} + \Omega_1 = \frac{p_2}{\rho} + \frac{q_2^2}{2} + \Omega_2 = C \quad \ldots (8.17)$$

where C is a constant. If motion occurs under the action of gravity only, $\Omega = gz$ where z is the vertical displacement, and

$$\frac{p}{\rho} + gz + \frac{q^2}{2} = C_1 \quad \ldots (8.18)$$

If external forces are neglected, then Bernoulli's theorem indicates that as the pressure is lowered the velocity of the fluid increases and

228

vice versa. If a liquid flows through a pipe having a constriction, then the velocity at the constricted part is increased and the pressure is thereby reduced. There are several practical applications of this effect.

8.4 ILLUSTRATIONS OF BERNOULLI'S THEOREM
A rapidly moving stream of water is passed through the inverted thistle funnel into the mouth of which a ball is placed [*Figure 8.3(a)*]. The water stream passing over that part of the ball close to the mouth

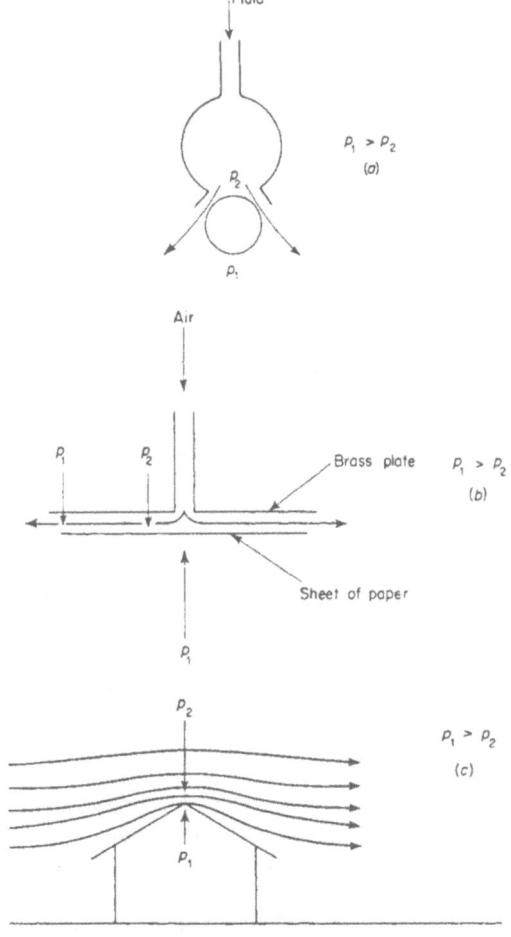

Figure 8.3. Illustrations of Bernoulli's theorem

of the funnel produces a low pressure within the fluid and hence, the thrust due to the atmospheric pressure holds the ball in position.

Figure 8.3(b) shows a brass plate, fitted with a centre tube, placed over a sheet of paper. If a stream of air is passed through the tube the air passes outwards and its velocity is greatest near the centre of the plate. Hence the air pressure is lowest near the centre and since at the edge it is approximately atmospheric, the paper does not fall.

Figure 8.3(c) illustrates a high wind blowing over the roof of a house. Due to the increased velocity over the apex of the roof, the pressure immediately above the roof is reduced. In a high wind, this reduction in pressure is considerable and the roof is lifted away. In the high winds experienced in America and other countries it is frequently found that the roof of a house is lifted off in this manner without damage to the rest of the structure.

8.5 THE VENTURI METER

This is an instrument for measuring the flow of fluid through a pipe. It was developed by Herschel in 1887 and is called after an Italian who carried out experiments on the flow of water in the eighteenth century. The Venturi meter is an excellent example of a practical application of Bernoulli's theorem. Basically it consists of two conical tubes joined by a short length of cylindrical tubing of relatively small diameter. In use, the meter is placed in a pipeline in a horizontal position. As the fluid flows through the meter its velocity will be a maximum in the constriction due to the reduction in cross-sectional area. Hence, the pressure will be reduced and this reduction in pressure is measured by the difference in the fluid levels as indicated in *Figure 8.4.*

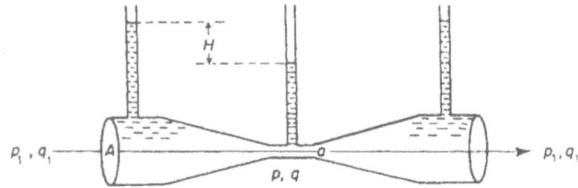

Figure 8.4. The Venturi meter

Let a be the cross-sectional area of the throat, and A that of the main pipeline. If friction, viscous forces, and eddy currents are neglected, then, by applying Bernoulli's theorem

$$\frac{q_1^2}{2} + \frac{p_1}{\rho} = \frac{q_2^2}{2} + \frac{p}{\rho} \qquad \ldots (8.19)$$

230

where p and q are the pressure and velocity at the constriction, p_1 and q_1 are the pressure and velocity in the main pipeline, and ρ is the fluid density. If Q is the volume of fluid passing per second, then

$$Q = q_1 A = qa \qquad \qquad \dots (8.20)$$

If the difference in fluid levels is H, then

$$p_1 - p = \rho g H \qquad \qquad \dots (8.21)$$

Hence, substituting for q_1, q, p_1 and p in equation (8.19)

$$\frac{Q^2}{2A^2} = \frac{Q^2}{2a^2} - gh$$

therefore

$$Q = \frac{aA}{(A^2 - a^2)^{\frac{1}{2}}} \cdot (2gH)^{\frac{1}{2}} \qquad \qquad \dots (8.22)$$

Thus a knowledge of a and A together with a measurement of H, gives the rate of flow of fluid through a pipe. In practice, viscous forces and eddy currents cannot be neglected and give a value of Q slightly greater than the theoretical value, so that

$$Q = \frac{KaA}{(A^2 - a^2)^{\frac{1}{2}}} \cdot (2gH)^{\frac{1}{2}} \qquad \qquad \dots (8.23)$$

where K is a constant for a given meter.

8.6 TORRICELLI'S THEOREM

Suppose that a fluid is flowing from a constant pressure tank through a long narrow pipe where the end Y is open to the atmosphere. Let the area of the free surface of the fluid be A_1 and let q_1 be its velocity (*Figure 8.5*). Let the depth of the pipe be z below the free surface and let the area of the orifice be A. Let the velocity of efflux

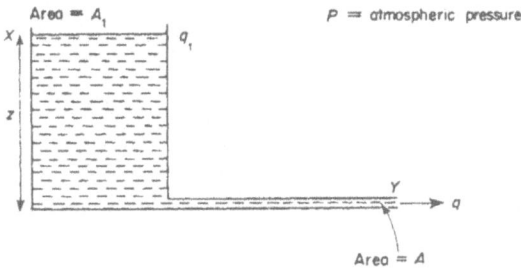

Figure 8.5. Torricelli's theorem

231

of the fluid be q and assume the atmospheric pressure to be p and the density of the fluid to be ρ.

If the motion of the fluid is assumed to be streamline then, neglecting viscous forces, etc., by applying Bernoulli's theorem to the flow at X and Y

$$gz + \frac{q_1^2}{2} + \frac{p}{\rho} = \frac{q^2}{2} + \frac{p}{\rho} \qquad \ldots (8.24)$$

i.e.
$$q^2 = q_1^2 + 2gz \qquad \ldots (8.25)$$

but $A_1 q_1 = Aq$, so that

$$q^2 = \frac{A_1^2 \cdot 2gz}{A_1^2 - A^2} \qquad \ldots (8.26)$$

When the diameter of the pipe is small compared with the diameter of the free surface, then A may be neglected in comparison with A_1, i.e. $q_1 \to 0$ and hence

$$q^2 = 2gz \qquad \ldots (8.27)$$

This relationship is known as Torricelli's theorem.

8.7 EFFLUX OF GASES

Let p_1 and ρ_1 be the pressure and density of a gas in a containing vessel. Let the gas flow through an orifice into a gaseous atmosphere, the pressure and density of the gas on emergence from the orifice being p_0 and ρ_0 respectively. Assume that there are no external forces acting and that adiabatic conditions prevail, i.e. $pv^\gamma = $ constant, where γ is the ratio of the specific heats of the gas. For steady motion of the gas Bernoulli's theorem may be applied, so that

$$-\int_{p_1}^{p_0} \frac{dp}{\rho} = \frac{q^2}{2} \qquad \ldots (8.28)$$

where q is the velocity of efflux of the gas. Now $p/\rho^\gamma = C$, where C is a constant. Hence

$$\frac{1}{\rho} = \frac{C^{1/\gamma}}{p^{1/\gamma}}$$

Therefore, from equation (8.28)

$$\frac{q^2}{2} = -\int_{p_1}^{p_0} \frac{C^{1/\gamma} \, dp}{p^{1/\gamma}}$$

232

Therefore $\quad q^2 = 2C^{1/\gamma} \cdot \dfrac{\gamma}{\gamma - 1} \left[p_1^{(\gamma - 1)/\gamma} - p_0^{(\gamma - 1)/\gamma} \right]$

but, $\qquad\qquad\qquad C^{1/\gamma} = \dfrac{p_1^{1/\gamma}}{\rho_1}$

Hence, $\qquad q^2 = \dfrac{2\gamma}{\gamma - 1} \cdot \dfrac{p_1}{\rho_1} \left[1 - \left(\dfrac{p_0}{p_1} \right)^{(\gamma - 1)/\gamma} \right]$ $\qquad$ .... (8.29)

This equation shows that an increase in velocity is accompanied by a decrease in pressure. A practical application of this is illustrated by the domestic vacuum cleaner.

8.8 VORTICITY

The vorticity is twice the angular velocity of an infinitesimally small fluid element, i.e. if a spherical fluid element could be instantaneously solidified and isolated from the remaining fluid it would rotate with an angular velocity equal to half the vorticity value. The vorticity vector w is defined as

$$w = \operatorname{curl} v \qquad\qquad .... (8.30)$$

where v is the velocity vector at the point in the fluid being considered. An equation for the vorticity may be derived as follows. Consider Euler's equation [equation (8.12)]. By applying the standard vector relation

$$(v \operatorname{grad}) v = [\operatorname{curl} v \times v] + \operatorname{grad} (q^2/2) \qquad .... (8.31)$$

where $q^2 = u^2 + v^2 + w^2$, equation (8.12) becomes

$$\frac{\partial v}{\partial t} - [\operatorname{curl} v \times v] + \operatorname{grad} (q^2/2) = -\frac{1}{\rho} \operatorname{grad} p + F \quad .... (8.32)$$

Taking the curl of each term in this equation and assuming F to be conservative

$$\frac{\partial w}{\partial t} + \operatorname{curl} [w \times v] = 0 \qquad\qquad .... (8.33)$$

By using the vector relation

$$\operatorname{curl} [w \times v] = (v \operatorname{grad}) w - (w \operatorname{grad}) v + w \operatorname{div} v \quad .... (8.34)$$

equation (8.33) becomes

$$\frac{\partial w}{\partial t} + (v \operatorname{grad}) w = (w \operatorname{grad}) v - w \operatorname{div} v \qquad .... (8.35)$$

233

But it can be shown that the total derivative

$$\frac{d}{dt} = \frac{\partial}{\partial t} + (v\ \text{grad}) \qquad\qquad \ldots\ (8.36)$$

Hence, equation (8.35) becomes,

$$\frac{dw}{dt} = (w\ \text{grad})\,v - w\ \text{div}\ v \qquad\qquad \ldots\ (8.37)$$

For an incompressible fluid, div $v = 0$, so that equation (8.37) becomes

$$\frac{dw}{dt} = (w\ \text{grad})\,v \qquad\qquad \ldots\ (8.38)$$

Equation (8.38) leads to an important result for incompressible fluid flow. If w is zero at any point on a streamline at any given time, then it is zero everywhere along the streamline, since, from equation (8.38), it may be seen that if $w = 0$ at any point on a streamline, then

$$\frac{dw}{dt} = 0 \qquad\qquad \ldots\ (8.39)$$

at all points along the streamline. It also follows from this result that for incompressible fluid flow, regions where the vorticity is zero are sharply separated from regions where the vorticity is not zero by a set of streamlines. This theoretical result can be modified by the effect of viscous forces, etc.

In the case where the space variation of the velocity is at right angles to the plane of the motion, i.e. to the vorticity, equation (8.39) applies over a wide range of conditions. In this flow situation all the lines of fluid motion are parallel to a fixed plane and the velocity vectors, at corresponding points, of all planes parallel to the reference plane are identical. This type of flow is referred to as two-dimensional incompressible motion and the vorticity is conserved in each fluid element as it moves in the fluid.

9

OSMOSIS, DIFFUSION AND BROWNIAN MOTION

9.1 OSMOSIS

IN 1748 Nollet showed that if an animal bladder was used to separate water and alcohol, water passed through the bladder into the alcohol but no alcohol passed through the bladder into the water. The bladder is said to possess a property of selective transmission and it may be considered to consist of a mesh of such a size that small particles such as molecules of water can pass through, while the molecules of alcohol are too large to penetrate. Such a membrane is termed a semi-permeable membrane and the term 'osmosis' is used to describe the preferential transmission of a liquid through such a membrane.

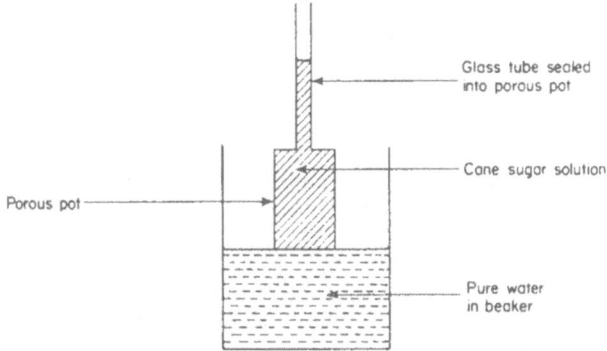

Figure 9.1. Osmosis

The process of osmosis can be simply demonstrated by the following experiment. A cylindrical porous pot has a semi-permeable membrane of cupric ferrocyanide deposited in its walls. It is then filled with a solution of cane sugar in water. A long glass tube is fixed through the top of the cylinder (*Figure 9.1*) and the cylinder is placed in a beaker of water. After some time the liquid inside the glass tube is found to have risen to a considerable height and thus water must have passed through the semi-permeable membrane into the sugar solution.

In actual fact, since it is an aqueous solution of sugar, some water molecules will also have passed from the sugar solution into the water in the beaker. The membrane is continually being bombarded by molecules of water on one side and by a mixture of water and sugar molecules on the other side. The sugar molecules are unable to pass through the membrane and thus there is an excess of water molecules entering the porous pot. As the pressure inside the porous pot increases due to the excess water entering it, it will enable more water molecules to leave the porous pot and ultimately a state of dynamic equilibrium is attained when as many water molecules are leaving the porous pot as are entering it. When this state is reached the height of the liquid in the glass tube remains constant and the excess pressure inside the vessel is termed the osmotic pressure. The osmotic pressure of a solution is defined as that pressure which must be applied to a solution to prevent the spontaneous differential flow of liquid through a semi-permeable membrane separating the solution and solvent.

9.2 THE LAWS OF OSMOSIS

Early quantitative work was carried out by Pfeffer who measured the osmotic pressure of a number of weak solutions. For non-electrolytic solutions, Pfeffer's experiments showed that the osmotic pressure, P, was directly proportional to the concentration of the solution, C, in moles per cubic centimetre i.e.

$$P \propto C \qquad \qquad \dots (9.1)$$

Pfeffer also showed that for a solution of fixed concentration the osmotic pressure was directly proportional to the absolute temperature, T, i.e.

$$P \propto T \qquad \qquad \dots (9.2)$$

In 1886 van't Hoff, after studying Pfeffer's results enunciated the two laws of osmosis.

(1) The osmotic pressure of a dilute solution is directly proportional to the concentration of the dissolved substance.

(2) The osmotic pressure of a dilute solution is directly proportional to its absolute temperature.

Combining these two laws and putting $C = 1/V$, where V is the volume of solution containing 1 mole of solute, then

$$PV = RT \qquad \qquad \dots (9.3)$$

where R is a constant. This formula is only applicable to dilute solutions. With electrolytic solutions the results obtained are

236

complicated by dissociation. Van't Hoff carried out an experiment to compare the osmotic pressure of a sucrose solution with the pressure of hydrogen at the same temperature and concentration and showed that the constant R in equation (9.3) is identical with the universal gas constant. Subsequently, van't Hoff also deduced this relationship thermodynamically. Theory and experiment are in good agreement where dilute solutions are concerned.

9.3 VAPOUR PRESSURE
It is found that the vapour pressure of a solution is less than that of the solvent alone. This difference may be accounted for as follows. *Figure 9.2* shows an inner vessel divided into two parts by a semipermeable membrane. Part A contains the solution of vapour pressure p_1, while part B contains the pure solvent of vapour pressure p. The vessel is contained within an outer vessel in which only the vapour, of density σ, is present, in addition to the liquids. Due to osmosis the liquid level in A will be higher than that in B.

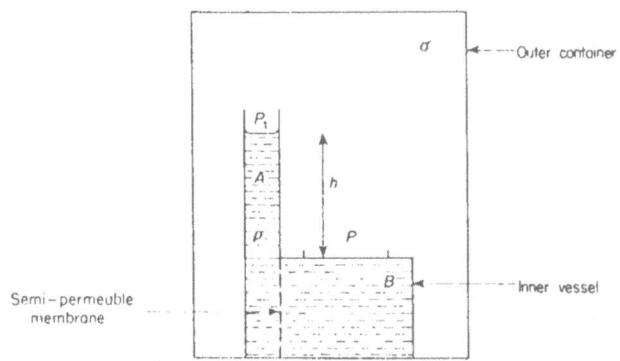

Figure 9.2. Vapour pressure of a solution

Let the height of the liquid in A over that in B be h and let the difference in pressure on the two sides of the membrane, i.e. the osmotic pressure of the solution, be P. Equating the osmotic pressure to the hydrostatic pressure difference gives

$$P = p_1 + \rho g h - p \qquad \dots (9.4)$$

where ρ is the density of the solution. Also p and p_1 are vapour pressures at a difference in height h in the vapour of constant density σ. Hence

$$P = p_1 + \sigma g h \qquad \dots (9.5)$$

237

I*

Combining equations (9.4) and (9.5) gives

$$p - p_1 = \frac{P\sigma}{\rho - \sigma} \qquad \dots (9.6)$$

σ is the vapour density under its own pressure and if σ_0 represents the density under standard atmospheric pressure P_0, then

$$\sigma = \sigma_0 p / P_0$$

$$\frac{p - p_1}{p} = \frac{P\sigma_0}{P_0(\rho - \sigma)} \qquad \dots (9.7)$$

In the derivation of the above equations the vapour density was assumed to be constant. This is only true for a short column of vapour. If this condition is not fulfilled then

$$\delta p = -g\sigma\delta h$$

Hence

$$\delta h = -\frac{P_0}{g\sigma_0} \cdot \frac{\delta p}{p}$$

On integration

$$h = \frac{P_0}{g\sigma_0} \log \frac{p}{p_1}$$

If it is then assumed that $P = \rho g h$, then

$$\log \frac{p}{p_1} = \frac{P\sigma_0}{P_0\rho} \qquad \dots (9.8)$$

Equation (9.8) provides a relationship between the osmotic pressure and the lowering of the vapour pressure of any solution. It is independent of any assumptions concerning the nature of osmotic pressure.

9.4 DETERMINATION OF MOLECULAR WEIGHTS

Let X be the volume of a solution containing x moles of dissolved substance, so that $V = X/x$. Hence

$$P(X/x) = RT$$

If the concentration, in moles per cubic centimetre, is C, then

$$P/C = RT$$

Now the concentration in grammes per cubic centimetre is $c = MC$, where M is the molecular weight of the substance, so that

$$P/c = RT/M \qquad \dots (9.9)$$

238

Hence, if the osmotic pressure, in absolute units, of a solution at a temperature T and concentration c, is measured, the molecular weight of the dissolved substance may be determined. Measurement of osmotic pressure is a standard technique in chemical laboratories for the determination of molecular weight. When a new compound, such as a polymer, has been prepared, osmotic pressure measurements afford a convenient method of determining the molecular weight.

9.5 MEASUREMENT OF OSMOTIC PRESSURE
9.5.1 Method of Berkeley and Hartley

Berkeley and Hartley measured the osmotic pressure of many solutions by a technique which they developed in 1906. Their apparatus is shown diagrammatically in *Figure 9.3*. It consists of a horizontal porcelain tube, about 15 cm long and 2 cm external diameter, with a semi-permeable membrane of copper ferrocyanide deposited near to its outer wall. A metal case encloses this tube and this case is filled with the solution whose osmotic pressure is required. The inlet to the case is connected to a hydraulic device for increasing the hydrostatic pressure on the solution, and also to a manometer. The ends of the porcelain tube are connected to capillary tubes.

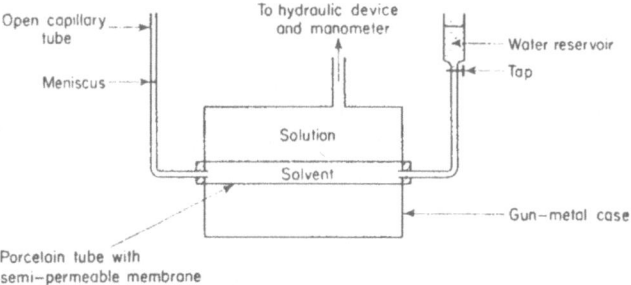

Figure 9.3. Apparatus of Berkeley and Hartley

Water is placed inside the porcelain tube and when the tube and capillary tubes are full of water, the tap on the inlet tube is closed, the open capillary tube serving as a water gauge. Water, or the solvent, tends to pass through the membrane but it is prevented from doing so by adjustment of the hydrostatic pressure on the solution. Adjustment of the applied hydrostatic pressure is continued until the meniscus in the open capillary tube remains stationary. If the applied pressure is less than the osmotic pressure, water passes from within the porcelain tube to the outer vessel and the meniscus drops. Similarly, if the applied pressure exceeds the osmotic pressure, the meniscus rises. When the meniscus is stationary the applied hydrostatic pressure is assumed to be equal to the osmotic pressure. The membrane used by Berkeley and Hartley allowed osmotic pressures of over 100 atm to be measured.

9.5.2 Work of Morse

Morse and his co-workers carried out a very thorough series of experiments on the measurement of osmotic pressure and ultimately developed a technique which enabled measurements to be carried out on concentrated solutions where high pressures are involved. The great difficulty in such measurements is the construction of a suitable semi-permeable membrane which must have uniform strength and porosity together with a fine texture. Morse finally used a mixture of clays to achieve these qualities. His apparatus is shown, in principle, in *Figure 9.4*. The semi-permeable membrane

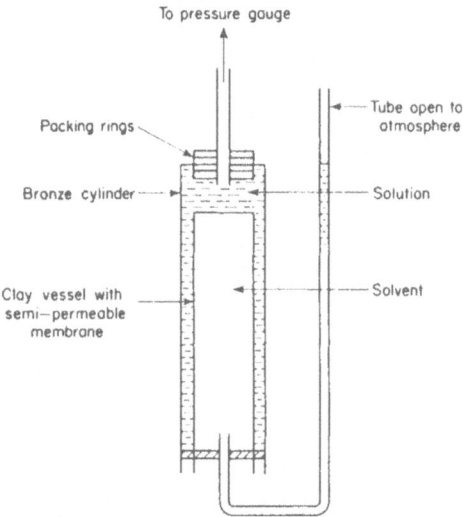

Figure 9.4. Principle of Morse's apparatus

of copper ferrocyanide is deposited on the inner wall of the clay vessel. This vessel is enclosed within a bronze cylinder. The solution under examination fills the space between the clay vessel and the bronze cylinder while the solvent, water, is contained within the clay vessel. The water is supplied through the tube which is open

240

to the atmosphere. In the apparatus used by Morse, leaks were overcome by the use of packing rings.

Water passes through the membrane into the solution and hence the pressure steadily increases in the solution until equilibrium is reached. The pressure was measured in the initial experiments by a manometer, but in later work, an electrical resistance gauge was used. The principle of operation of such gauges has been described in Sub-section 5.28 dealing with Bridgman's work at high pressures.

9.5.3 The osmotic balance

Most modern instruments for measuring osmotic pressure utilize a pressure gauge or manometer to measure the pressure difference across the membrane. However, for accurate work, the osmotic balance is sometimes used. This consists of a balance with one of the scale pans replaced by an osmotic cell, suspended from the balance arm, in the base of which is the semi-permeable membrane. The cell is filled with the solution under examination and is placed in a glass cylinder filled with water. This cylinder is on an adjustable platform so that initially the meniscus in the cell can be arranged to be level with the meniscus in the cylinder. The balance is first adjusted until balance is attained. Due to osmosis the solvent flows into the cell thus causing it to sink in the cylinder until the buoyancy compensates for the increase in weight. Readjustment of the balance to equilibrium gives the additional weight necessary and from this the osmotic pressure may be calculated.

9.6 OSMOTIC PRESSURE THEORY

While the results of various workers using different techniques, agree remarkably well, there is a marked disagreement between experiment and theory. The formula $PV = RT$ is only applicable to dilute solutions of non-electrolytes, so when more concentrated solutions are examined the results no longer fit this formula. The classical theory of osmotic pressure of dilute solutions assumes that no forces exist between the molecules or ions, and that osmotic pressure is similar to the pressure of gases and is calculated as the normal momentum imparted to the boundary surface, per square centimetre per second by the impact of the molecules or ions. Just as the kinetic theory of gases was modified by van der Waals so the theory of osmotic pressure was modified by Debye and Huckel's theory of strong electrolytes. Strong electrolytes, in this context, are substances which are completely dissociated into ions in dilute, and even in moderately strong, solutions. Weak electrolytes, on the other hand, when dissolved in water, give solutions in which dissociation into ions takes place only to a very small extent and their behaviour approximates to that of classical theory.

Debye and Huckel's theory takes into account the forces exerted by the ions on each other due to their electric charges. The forces are given by the law of inverse squares and are dependent on the value of the dielectric constant of the solvent which fills the spaces between the ions. Since any particular ion is more likely to be approached by ions of opposite charge, the kinetic energy of the ions is affected and the osmotic pressure is reduced. The charged ions also exert forces on each other in weak electrolyte solutions but since there are very few of them compared with the large number of undissociated molecules, the effect of the forces is very small and so the behaviour of weak electrolytes is as predicted by classical theory.

9.7 DIFFUSION IN LIQUIDS

If two liquids in contact are able to mix in any proportions they do so spontaneously until a uniform mixture is produced. This process is called diffusion and it is due to the wandering of the liquid molecules or ions in the solution, from a region of high concentration to one of low concentration. The process resembles the conduction of heat in a metal from a region of high temperature to one of low temperature. Diffusion takes place rapidly in gases and relatively slowly in

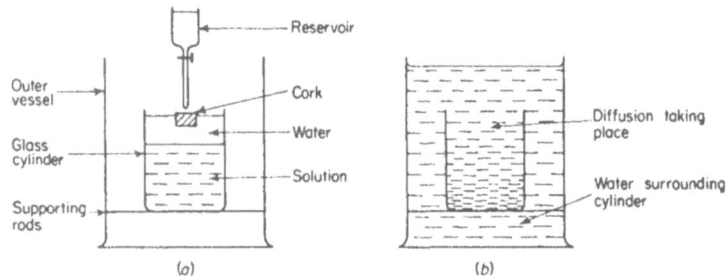

Figure 9.5. Diffusion of liquids

liquids, the rapidity of movement being dependent on the rate of change of density at the place considered. It is an irreversible process, i.e. once pure solvent is impregnated with a solute it cannot return to its pure state unless aided by an external agent.

The diffusion of liquids was first studied in 1850 by Graham who used apparatus similar to that shown in *Figure 9.5(a)*. A small glass cylinder rests inside a larger vessel, on horizontal supporting rods. The solution being examined is placed in the cylinder and a cork is floated on the liquid surface. The cork is arranged to remain in the centre of the solution surface. Water is run from the funnel and allowed to fall on the wetted cork at a steady rate of two or three drops per second. In this way a layer of water is gradually formed on top of the solution. When the cork is finally free of the solution it is removed and the cylinder completely filled with the water. The outer vessel is then filled with water as in *Figure 9.5(b)* and the temperature maintained constant in order to avoid convection currents.

Initially the boundary between the solution and water is quite sharp and clear but as diffusion takes place this distinct boundary gradually disappears. Graham determined the amount of solute which diffused from the inner vessel into the water and his experiments showed that the rate of diffusion of aqueous solutions depended on the type of salt used, was directly proportional to the concentration of the dissolved salt and increased with an increase in temperature.

9.8' FICK'S LAW

A few years after Graham's experiments on diffusion, Fick summarized his results in a simple mathematical formula. His formula states that the mass, m, of a substance in solution passing across an area, A, per second, is directly proportional to the rate at which the concentration C, of the dissolved substance diminishes in a direction at right angles to the plane of A, i.e.

$$\frac{m}{A} = -k\frac{\partial C}{\partial x} \qquad \dots (9.10)$$

where $\partial C/\partial x$ is the concentration gradient or change of concentration with distance, and k is a constant called the 'coefficient of diffusion' of the dissolved substance.

If two parallel planes in the liquid are considered, each of unit area and at a distance δx apart, then, if the concentration at one of the planes at a time t is C, the concentration at the other will be $C - (\partial C/\partial x)\,\delta x$. The inflow of dissolved substance at the first plane in time δt is given by $k(\partial C/\partial x)\,\delta t$ and the outflow from the other plane is

$$k\frac{\partial C}{\partial x}\,\delta t - k\frac{\partial^2 C}{\partial x^2}\cdot\delta x\delta t$$

243

Hence the volume between the two planes has a net gain of $k(\partial^2 C/\partial x^2) \cdot \delta x \delta t$. The volume is δx, and the change of concentration is $k(\partial^2 C/\partial x^2) \cdot \delta t$. Thus the rate of change of concentration, $\partial C/\partial t$, is given by the equation

$$\frac{\partial C}{\partial t} = k \frac{\partial^2 C}{\partial x^2} \qquad \cdots (9.11)$$

Equation (9.11) is the general equation governing the diffusion process. The equation assumes that the liquid is at rest during the diffusion process. In experiments on diffusion some movement of the liquid always takes place so that it is essential to modify equation (9.11) when comparing theory and experiment.

Fick attempted to verify his law by carrying out an experiment when the diffusion had attained a steady state, i.e. when $\partial^2 C/\partial x^2 = 0$. He placed a tube, open at both ends, vertically in a vessel containing salt crystals so that the lower end of the tube was always in contact with the crystals. The tube was filled with water and the entire assembly was placed in a large tank of water and left until the steady state had been attained. The concentration at the lower end of the tube is constant and it may be assumed to be zero at the top of the tube. Since $\partial^2 C/\partial x^2 = 0$, the concentration is given by

$$C = ax + b \qquad \cdots (9.12)$$

where a and b are constants. Fick took a small glass bulb and weighed it carefully at various depths below the liquid surface within the tube. This enabled him to calculate the density of the solution at varying depths. Hence, from a knowledge of the relationship between density and concentration, Fick determined the concentration. His results were in good agreement with equation (9.12).

9.9 MEASUREMENT OF THE COEFFICIENT OF DIFFUSION BY CLACK'S METHOD

Clack measured the coefficient of diffusion k for aqueous solutions of sodium chloride, potassium chloride and potassium nitrate of various strengths. His method is based on Fick's law, equation (9.10). Consider the upward diffusion of a solute in a uniform vertical tube (*Figure 9.6*). At the bottom of this tube the solution is kept saturated by the presence of crystals of solute. At the top of the tube a slow, steady stream of water flows across and by carrying off the solute arriving there it becomes a very dilute solution.

Let the concentration of the solution in grammes per cubic centimetre at M, distant x cm from the bottom of the tube and 1 cm from the top, be C and let ρ be the density of the solution at this point. Let i be the net decrease in mass of the system, in grammes per second, under steady state conditions, due to the departure of solute and arrival of water at the top. Let δ be the ratio of the mass

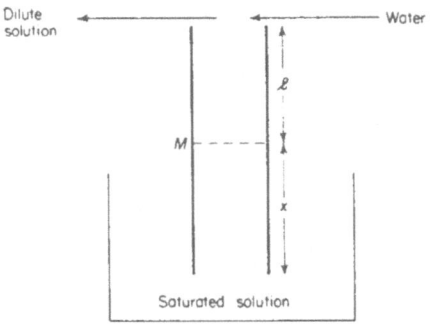

Figure 9.6. Measurement of coefficient of diffusion

of water entering the top of the cell per second to the mass of salt leaving per second. Equation (9.10) can be rearranged as

$$k = \frac{m}{A}\frac{\partial x}{\partial C} \qquad \dots (9.13)$$

and by a consideration of the dynamic equilibrium of the above arrangement this equation can be put into such a form that direct substitution from experimental results can be made. In fact, Clack showed that

$$k = \frac{i}{A(1-\delta)}\cdot\frac{dl}{dC}\cdot\frac{\rho - C - C\delta}{\rho - C} \qquad \dots (9.14)$$

dl/dC may be split up into two factors, $dl/d\mu$ and $d\mu/dC$ where μ is the optical refractive index of the solution at the level M. Both these factors can be determined experimentally.

Figure 9.7 illustrates the experimental arrangement used by Clack. The vertical tube is about 5 cm long with a cross-section of roughly 1 cm by 4 cm. It fits into a glass box which is filled with the saturated solution and crystals of the solute being investigated. This system is contained in a glass cell filled with distilled water. An inlet tube, through which distilled water enters, and an outlet tube by which very dilute solution leaves, are fitted as shown. If this system is left for several days it attains a steady state to which equation (9.14) is applicable. The whole apparatus must be maintained at a constant temperature throughout the experiment.

The change in weight per second under steady state conditions, i, may be directly determined by initially suspending the cell from the arm of a balance. To determine $d\mu/dl$, the rate of change of the refractive index of the liquid with depth, Clack measured the deviation of an initially horizontal ray of light as it traversed the cell in a

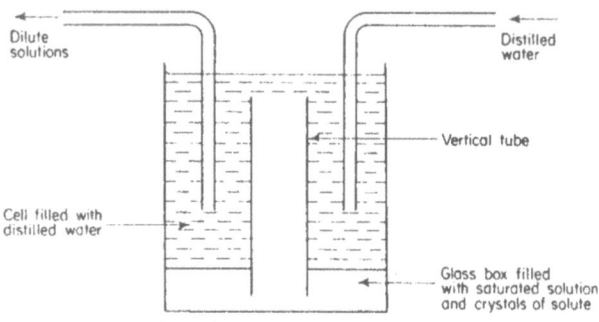

Figure 9.7. Clack's cell

direction perpendicular to the plane of *Figure 9.7*. The ray is refracted in a vertical plane and follows a curved path, finally emerging in a downward direction making an angle α with its original direction. It can easily be shown that

$$\frac{d\mu}{dl} = \frac{\sin \alpha}{d}$$

where d is the horizontal thickness of the cell. Clack determined $\sin \alpha$ by measuring the vertical displacement of the central fringe in the interference pattern produced by two narrow and close horizontal slits illuminated by green light from a mercury arc. The optical arrangement is shown in *Figure 9.8*, from which it can be seen that the green light emerges from a horizontal slit and is made parallel and horizontal by a lens. After passing through two horizontal slits in the vertical screen it traverses the cell and is finally caused to converge on the eyepiece by another lens. The two slits can be moved in a vertical direction so that the entire height of the cell can be examined. To eliminate the effect of imperfections in the cell the readings are repeated with the cell full of distilled water.

The change of refractive index with concentration of solution, $d\mu/dC$, is determined by the use of a Rayleigh refractometer and hence the change of concentration with depth, dl/dC, may be found. If the coefficient of diffusion were independent of the concentration gradient in the solution, then, under steady conditions, the concentration would be constant in Clack's experiment. However, the results of the experiment showed a varying concentration gradient under steady conditions and the variation of the coefficient of diffusion, k, with concentration, C, was found. For each of the solutions used, Clack showed that as the concentration increased steadily, the coefficient of diffusion first reaches a minimum value and then increases almost linearly.

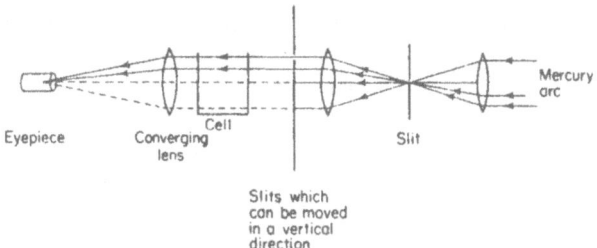

Figure 9.8. *Optical arrangement*

9.10 DIFFUSION AND OSMOTIC PRESSURE

The osmotic pressure and the diffusion of a solute in solution are connected. If the concentration of the solution at one point is greater than that at another point then, according to van't Hoff's law, the osmotic pressure is also greater. The osmotic pressure may be regarded as a force giving the molecules of the solution an acceleration from one point to another point at a lower concentration.

Consider a cylinder of cross-section A, containing a non-electrolytic solution in which diffusion is taking place. Let the osmotic pressure at a position x in the cylinder be P and at the point $x + \delta x$, let it be $P + (dP/dx) \, \delta x$ as shown in *Figure 9.9*. The net force due to osmotic pressure, acting on the cylinder of length δx in the direction of increasing x, is thus $(-dP/dx) \, A\delta x$. Let C be the concentration of the solution in moles per cubic centimetre, so that the number of moles in the volume $A\delta x$ is $CA\delta x$. Hence the force acting in the x direction on each mole in the layer is

$$\frac{-A\delta x}{CA\delta x}\left(\frac{dP}{dx}\right) = -\frac{1}{C}\frac{dP}{dx} \qquad \ldots (9.15)$$

247

If the motion of the molecules is subject to retarding forces such as viscosity, the molecules acquire a constant terminal velocity. Let the retarding force on each mole when it is moving with a constant velocity of 1 cm/s be F. F is also equal to the force necessary to drive 1 mole through the solution with a velocity of 1 cm/s. Thus the force $(-1/C \cdot dP/dx)$ produces a terminal velocity of $(-1/CF \cdot dP/dx)$. If the number of moles which pass across each layer in unit time is dN/dt, then

$$\frac{dN}{dt} = -\frac{1}{CF}\frac{dP}{dx} \cdot AC = -\frac{A}{F}\frac{dP}{dx}$$

but for dilute solutions, by van't Hoff's law $P = CRT$ and hence

$$\frac{dN}{dt} = -\frac{RT}{F} \cdot A \frac{dC}{dx} \qquad \ldots (9.16)$$

Also by Fick's law

$$\frac{dN}{dt} = -kA\frac{dC}{dx} \qquad \ldots (9.17)$$

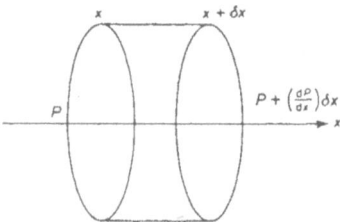

Figure 9.9 Diffusion and osmotic pressure

Thus, the coefficient of diffusion, k, corresponds to the factor RT/F. From equations (9.16) and (9.17) the force necessary to drive 1 mole through the solution with a velocity of 1 cm/s can be calculated, if k is known, i.e.

$$F = -\frac{RT}{dN} \cdot A \cdot \frac{dC}{dx} \cdot dt = \frac{RT}{k} \qquad \ldots (9.18)$$

Since diffusion in liquids is a very slow process the practical unit of time used is the day.

9.11 DIFFUSION OF ELECTROLYTES
Suppose a solution of a single electrolyte contains two univalent ions and let the velocities of the cations and anions when subjected

to unit force be u cm/s and v cm/s, respectively. As outlined in the previous section each ion is driven along by the osmotic pressure and the velocity of the cation is $-(u/C).(dP/dx)$ while that of the anion is $-(v/C).(dP/dx)$. The amounts of each ion passing any cross-section of the cylinder in unit time are, therefore $-uA(dP/dx)$ and $-vA(dP/dx)$. Usually u is not equal to v so that an electric field is set up and this field exerts a force on the ions.

If the potential at a point is E then the force on a gramme equivalent of an ion carrying a charge e is edE/dx. A force edE/dx produces a terminal velocity of $-ue\,dE/dx$ cm/s for the cations and similarly a terminal velocity of $+ve\,dE/dx$ cm/s for the anions. Hence the total terminal velocities of the cations and anions, due to both osmotic pressure and electric field are, respectively

$$-u\left(\frac{1}{C}\frac{dP}{dx} + e\frac{dE}{dx}\right) \text{ and } -v\left(\frac{1}{C}\frac{dP}{dx} - e\frac{dE}{dx}\right)$$

Now the total number of gramme equivalents of an ion crossing any cross-section of the cylinder in unit time under the influence of both osmotic and electric forces must be equal. If this quantity is dN/dt, then

$$\frac{dN}{dt} = -uA\left(\frac{dP}{dx} + Ce\frac{dE}{dx}\right) = -vA\left(\frac{dP}{dx} + Ce\frac{dE}{dx}\right) \quad \ldots (9.19)$$

Eliminating dE/dx gives

$$\frac{dN}{dt} = \frac{2uv}{u+v}.A.\frac{dP}{dx} \quad \ldots (9.20)$$

Now $\dfrac{dN}{dt} = -kA\dfrac{dC}{dx}$ {equation (9·17)} and $P = CRT$, hence

$$k = \frac{2uv}{u+v}.RT \quad \ldots (9.21)$$

Therefore, if u and v are known, the coefficient of diffusion, k, may be determined. The ion velocities can be calculated from a study of ion migration, and the resulting values of k are found to agree well with measured values.

9.12 BROWNIAN MOVEMENT
In 1827, the English botanist, Brown, was observing small pollen grains suspended in water through the, then new, achromatic objective of his microscope. He noticed that the particles were

continually moving in all directions, apparently quite spontaneously. Brown subsequently observed that all particles, if small enough, exhibited the same kind of motion and he ascribed this motion to so called 'vital forces'. It was almost 50 years after its discovery before an explanation for Brownian motion was suggested, first by Wiener and later, in 1877, by Delsaulx.

At this time the kinetic theory of matter had been well established by Clausius and Maxwell and it was suggested that Brownian movement was due to the thermal agitation of the molecules of the surrounding liquid. In a small time interval a particle receives more impacts on one side than on the other due to molecular bombardment and, if it is sufficiently small, it will move in the direction of the excess force until its path is changed by subsequent molecular impacts. Hence the observed random path of a small particle is the resultant of a large number of individual paths which are too small to be resolved by the microscope. Since the liquid molecules always possess thermal energy and are in motion, Brownian motion is continuous and everlasting. In fact it has been exhibited by small particles held in suspension in liquid-filled cavities in granite and in other very old rocks.

9.13 SEDIMENTATION EQUILIBRIUM
If a suspension or a colloidal solution is contained in a vessel, the suspended particles are subjected to the opposing forces of gravity

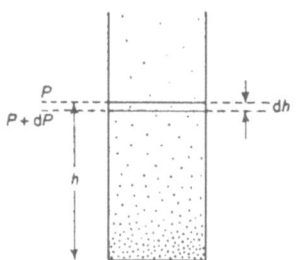

Figure 9.10. Sedimentation equilibrium

and osmotic pressure, which by kinetic theory, is due to the bombardment of the molecules. Ultimately a state of dynamic equilibrium is attained and a definite vertical concentration gradient is established, as shown in *Figure 9.10*.

Let the number of suspended particles per cubic centimetre, at a height h above the bottom of the vessel, be n and let the osmotic pressure at this height be P. Let the osmotic pressure at a height

$h - dh$, due to the increase in the number of particles, be $P + dP$. Now the increase in osmotic pressure, dP at this height is sufficient to counteract the weight of the particles in the layer dh, and hence

$$dP = nmg\, dh \qquad \ldots (9.22)$$

where m is the mass of a particle, after being corrected for buoyancy, and is given by

$$m = V_c(\rho - \rho_0) \qquad \ldots (9.23)$$

where V_c is the volume of one particle, ρ is the particle density and ρ_0 is the liquid density.

The osmotic pressure $P = CRT = nRT/N_A$ where N_A is Avogadro's number, i.e. $6 \cdot 02 \times 10^{23}$ molecules per mole. Hence

$$dP = RT\, dn/N_A \qquad \ldots (9.24)$$

Combining equations (9·22), (9·23) and (9·24) gives

$$\frac{RT}{N_A} dn = ng\, V_c(\rho - \rho_0)\, dh \qquad \ldots (9.25)$$

Rearranging and integrating

$$\log e \frac{n_1}{n_2} = \frac{N_A}{RT} V_c g\, (\rho - \rho_0)(h_1 - h_2) \qquad \ldots (9.26)$$

where n_1 and n_2 are the numbers of particles per cubic centimetre at heights h_1 and h_2, respectively, above the bottom of the vessel.

9.14 PERRIN'S EXPERIMENTS

In the early 1900's Perrin carried out what are now regarded as a classical series of experiments to test the validity of equation (9.26). Perrin used emulsions of gamboge or mastic and by careful preparation obtained the particles of uniform size essential to success. He first treated the gamboge with alcohol which dissolved the large amount of yellow matter present. Addition of water to this solution caused the formation of a yellow emulsion made up of very small spherical particles. This emulsion was then subjected to continuous centrifuging for several weeks and as a result, the spherical particles were separated into layers containing particles of the same size. A drop of the emulsion containing particles of a particular size was placed in a small cell formed by using a hollow microscope slide as shown in *Figure 9.11*, and these particles were observed through a microscope. The field of view was limited by a plate with a fine hole, so that only a small number of particles were visible at once.

The entire apparatus was maintained at a constant temperature to prevent convection currents being set up in the liquid.

With the microscope focused on one particular layer of the emulsion the number of particles visible in the field of view was counted. The microscope was then moved a known distance so that it was focused on another layer of the emulsion and again the number of particles present was counted. Several readings were taken at intervals for each position of the microscope so as to eliminate statistical errors. In this way the number of particles present at various heights in the emulsion was found.

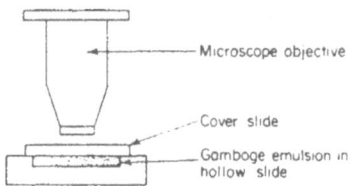

Figure 9.11. Perrin's apparatus

Perrin then determined the volume and density of the particles by various methods in order to check his values. One method of determining the density was by surrounding the particles with potassium bromide of steadily increasing concentration until a stage was reached where the particles neither floated nor sank. This occurred when the density of the particles and the solution was the same. The volume of the particles was determined by letting the emulsion almost dry on a glass plate so that the particles were pulled into rows by surface tension. The length of a row and the number of particles in it were then measured by a travelling microscope, from which the average radius of the particles was determined and hence their volume. The result was checked by a second experiment involving Stoke's law. A cloud of particles was formed by stirring the emulsion, and the terminal velocity determined as the particles descended under the influence of gravity. This was very small, of the order of a few millimetres per day. The value of the particle radius was determined from Stoke's law {equation (7.21)} in the usual way.

From his experimental results Perrin verified the validity of equation (9.26). He obtained a value for Avogadro's number, of approximately 6.8×10^{23} molecules per mole which is rather high. However, in 1915 Westgren, using the same technique, but working with colloidal particles of gold and silver, obtained a value for N_A of 6.04×10^{23} which is close to the now accepted value of 6.024×10^{23}.

9.15 EINSTEIN AND SMOLUCHOWSKI'S EQUATION

It has been established that Brownian motion is due to the particles being subjected to molecular bombardment. If the assumption is made that the mean kinetic energy of a suspended particle is the same as that of a gas molecule at the same temperature, an equation can be derived connecting the mean free path of the particle with the constants of the gas equation. The underlying theory was first proposed by Einstein in 1905 and later extended by Smoluchowski. Their derivation was rather complicated, however, and a few years later Langevin provided a simpler approach which is outlined below.

Let the average kinetic energy of each molecule be E, then

$$E = \frac{1}{2} mv^2 = \frac{3}{2} \frac{RT}{N_A} \qquad \ldots (9.27)$$

since from kinetic theory

$$PV = RT = \tfrac{1}{3} N_A mv^2$$

(the symbols having their usual meaning). Applying the principle of equipartition of energy, the kinetic energy due to motion in the x direction is $E/3$ and similarly in the y and z directions. The equation of motion of a particle in the x direction is thus

$$m\ddot{x} + \delta\dot{x} + X = 0 \qquad \ldots (9.28)$$

where X represents the force produced by molecular bombardment and δ represents the damping coefficient due to viscous forces. Multiplying equation (9.28) by x and putting

$$x\ddot{x} = \tfrac{1}{2}(\ddot{x}^2) - \dot{x}^2$$

one has

$$\frac{1}{2} m(\ddot{x}^2) - m\dot{x}^2 + \frac{\delta}{2}(\dot{x}^2) + Xx = 0 \qquad \ldots (9.29)$$

Now if this equation is applied to a large number of particles of the same size and the mean result considered, then the force X will be negative as often as it is positive and the average value of Xx is zero. Also the average value of $m\dot{x}^2$ is $2E/3 = RT/N_A$. Substituting in equation (9.29) gives

$$\frac{1}{2} m\dot{\alpha} - \frac{RT}{N_A} + \frac{\delta\alpha}{2} = 0 \qquad \ldots (9.30)$$

where α represents the mean value of $(\dot{x}^2)$. Integration of equation (9.30) gives

$$\alpha = \frac{2RT}{N_A \delta} + Ae^{-t\delta/m} \qquad \dots (9.31)$$

As t increases, $Ae^{-t\delta/m}$ rapidly approaches zero, even for gases, so that

$$\alpha = \frac{2RT}{N_A \delta} \qquad \dots (9.32)$$

Applying Stoke's law, $\delta = 6\pi\eta a$, where a is the particle radius. Hence

$$\alpha = (\dot{x}^2) = \frac{2RT}{6\pi\eta a N_A} \qquad \dots (9.33)$$

Integration of equation (9.33) gives

$$\overline{x^2} = \frac{RT}{3\pi\eta a N_A} \cdot t \qquad \dots (9.34)$$

where $\overline{x^2}$ denotes the mean value of the squared displacements for all the particles considered in the time interval t. Alternatively, equation (9.34) gives the average of the squared displacements for the same particle observed through several intervals of time. Smoluchowski pointed out that experimental results probably would not be in exact agreement with equation (9.34) since it had been assumed that all the particles were spherical and that they did not mutually attract each other.

Perrin carried out some experiments to test the validity of equation (9.34). He measured the displacements of a large number of particles in an emulsion, using a microscope fitted with a transparent squared grating in the eyepiece. Perrin's results confirmed that $\overline{x^2}$ is proportional to t and he obtained a value of 6.82×10^{23} for Avogadro's number.

9.16 BROWNIAN MOTION IN GASES

In 1909 Ehrenhaft reported the first results on the Brownian motion of particles suspended in gases. He found that there was a much greater activity in gases than in liquids as predicted by theory. Subsequent experiments were carried out by de Broglie but undoubtedly the most important experiments on Brownian motion in gases are those carried out by Millikan in 1911.

The principle of his experiment is the observation of the movement of a charged oil drop under the action of gravity alone and also when an electric field opposes the gravitational force. Millikan formed very small oil drops by blowing oil through an atomizer into the space between the plates of a parallel-plate condenser. Subsequent movement of the illuminated drops was observed through a microscope. The rate of fall of an oil drop under the action of gravity alone was observed and, applying Stoke's law, there results

$$6\pi\eta a v_1 = \tfrac{4}{3}\pi a^3(\sigma - \rho)g \qquad \qquad \ldots (9.35)$$

so that

$$v_1 = \frac{2ga^2}{9\eta}(\sigma - \rho) \qquad \qquad \ldots (9.36)$$

where, v_1 is the gravitational drift velocity, η is the viscosity of air, a is the drop radius, σ is the oil-drop density and ρ is the air density.

Application of an electric field of E e.s.u., acting against gravity, caused the drop to ascend with velocity v_2 and this velocity was also measured. If the effective mass of the drop is m and it possesses a charge e, then, since the terminal velocities of the drop are proportional to the forces acting

$$\frac{v_2}{v_1} = \frac{Ee - mg}{mg} \qquad \qquad \ldots (9.37)$$

Hence

$$e = \frac{mg(v_1 + v_2)}{Ev_1} \qquad \qquad \ldots (9.38)$$

But, from equation (9.35)

$$mg = 6\pi\eta a v_1$$

Hence, substituting in equation (9.38)

$$e = \frac{6\pi\eta a}{E}(v_1 + v_2) \qquad \qquad \ldots (9.39)$$

If this equation is combined with equation (9.34) there results

$$\overline{x^2} = \frac{2RT}{N_A Ee}(v_1 + v_2)t \qquad \qquad \ldots (9.40)$$

and hence the product $N_A e$ can be obtained without the drop radius or the viscosity of the air being measured. Millikan measured the lateral Brownian displacements of the drop when it was suspended in the gas so that the applied electric field just balanced the weight of the drop. Millikan worked with different gases at a low

255

pressure between the plates of the condenser so that the values of $\overline{x^2}$ were much greater than those observed by Perrin in the case of liquids. As a result of his experiments Millikan concluded that the value of $N_A e$ was $2 \cdot 88 \times 10^{14}$ e.s.u.

9.17 BRILLOUIN'S DIFFUSION EXPERIMENTS

Let n_1 and n_2 represent the numbers of particles per unit volume in an emulsion, at a distance apart equal to the root mean square displacement, $(\overline{x})$. Then, in a time t, the net interchange of particles between the two volumes is approximately

$$n = \frac{\overline{x}}{2}(n_1 - n_2) \qquad \dots (9.41)$$

But, from the definition of the coefficient of diffusion k

$$n = \frac{k(n_1 - n_2) t}{\overline{x}} \qquad \dots (9.42)$$

Hence $$\overline{x^2} = 2kt \qquad \dots (9.43)$$

and substituting for $\overline{x^2}$ in equation (9.34) gives

$$N_A = \frac{RT}{6k\pi\eta a} \qquad \dots (9.44)$$

Brillouin carried out experiments to verify equations (9.43) and (9.44). A glass plate, which acted as a perfect absorber for any particles coming into contact with it, was immersed in a suspension of gamboge in glycerine. If the average number of particles per unit volume of the emulsion is n', then the number of particles coming into contact with unit area of the plate in a time t is

$$n = \frac{n'\overline{x}}{2} \qquad \dots (9.45)$$

Thus, combining equations (9.43) and (9.45) gives

$$n^2 = \frac{n'^2 kt}{2} \qquad \dots (9.46)$$

Brillouin took photographs of the glass plate at regular intervals of time and verified that the number of particles collected was proportional to the time. He also obtained a value of $6 \cdot 9 \times 10^{23}$ for Avogadro's number.

10

PRODUCTION AND MEASUREMENT OF LOW PRESSURE

10.1 INTRODUCTION

THE production of low pressures was first investigated seriously in the late nineteenth century when the carbon filament electric light was developed. Early mechanical pumps made it possible to reach pressures of about 0·25 mm Hg but these pumps were superseded in the twentieth century by the rotary mercury pump which could attain a pressure of 10^{-5} mm Hg. In later years diffusion pumps, of various types, and molecular pumps were developed so that today, it is a routine procedure to obtain a pressure lower than 10^{-7} mm Hg or 10^{-7} torr, the torr being the unit normally used for expressing pressure in vacuum technology.

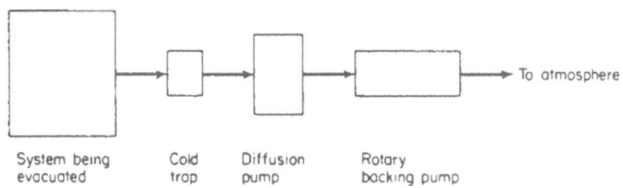

Figure 10.1. High vacuum system

A modern vacuum system normally consists of a backing pump arranged in series with a diffusion–condensation pump or a molecular drag pump, as shown in *Figure 10.1*. The backing pump is usually a rotary vane, oil-type pump, and its function is to reduce the pressure in the system from atmospheric to about 10^{-1} torr. At this pressure a diffusion pump, which will produce a pressure of 10^{-6} torr or better, can be used. If a mercury-vapour diffusion pump is used, a cold trap of liquid air is inserted to prevent the mercury from entering the system being evacuated.

The maintenance of a high vacuum is dependent on the system being free from leaks. The use of special high vacuum seals and joints assists in the construction of a leak-proof system but even so, the elimination of all leaks is often a painstaking procedure. The speed with which the degree of high vacuum required is established in a system depends on the pumps used and on the length and diameter of the connecting tubes. It can be shown from kinetic theory that, at low pressures, the rate of flow of gas is inversely proportional to the length and directly proportional to the cube of the radius of the tube so that, as far as possible, connecting tubes in a vacuum system should be wide and short.

The speed of a pump is measured by the rate at which the pressure is reduced in a given volume, and it is defined by the equation

$$\frac{dP}{dt} = -\frac{S}{V}(P - P_0) \qquad \qquad \ldots(10.1)$$

where S is the pumping speed at a pressure P, V is the given volume and P_0 is the limiting pressure attainable in the system. Integration and rearrangement of equation (10.1) gives the pumping speed as

$$S = \frac{V}{t_2 - t_1} \cdot \log\left(\frac{P_1 - P_0}{P_2 - P_0}\right) \qquad \ldots(10.2)$$

where P_1 and P_2 are the pressures in the given volume at the times t_1 and t_2. If the limiting pressure attainable in the system is very small indeed, then equation (10.2) can be simplified to

$$S = \frac{V}{t_2 - t_1} \cdot \log\frac{P_1}{P_2} \qquad \qquad \ldots(10.3)$$

Equation (10.3) is known as Gaede's equation and it is commonly used by manufacturers when providing data on vacuum pumps.

From equation (10.1), if the limiting pressure is neglected

$$S = -\frac{V}{P} \cdot \frac{dP}{dt} \qquad \qquad \ldots(10.4)$$

Now if the volume of gas removed from the volume V at a pressure P in time dt is dV

$$PV = (P + dP)(V + dV)$$

from which

$$-\frac{V}{P} \cdot \frac{dP}{dt} = \frac{dV}{dt}$$

Hence

$$S = \frac{dV}{dt} \qquad \qquad \ldots(10.5)$$

i.e. the speed of a pump is equal to the rate of change of volume of the gas in the system at any instant of time, the volume being measured at the pressure reached by the pump at that instant. This implies that at the lowest attainable pressure a pump has no speed.

10.2 TYPES OF PUMP

10.2.1 Rotary vacuum pumps

Mechanical pumps enable one to attain pressures of the order of 10^{-2} torr and they are generally used as backing pumps in conjunction with a diffusion pump. They can be of either the piston or the rotary type but whereas, at one time, piston pumps were in regular use, nowadays, almost all mechanical pumps are of the rotary type.

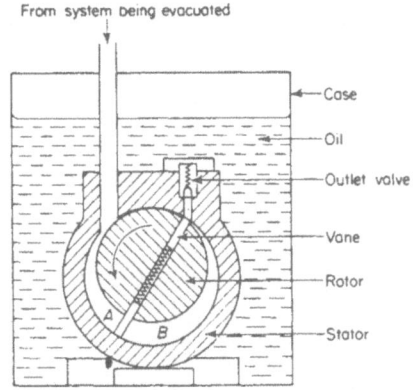

Figure 10.2. Gaede's rotary-vane oil pump

Gaede first suggested the rotary-vane mechanical pump and *Figure 10.2* illustrates a cross-section through this pump. It consists of a steel housing called the stator which has a cylindrical chamber bored in it. A solid cylindrical rotor is mounted so that its axis is parallel to, but eccentric from, the axis of the chamber. The rotor rotates about its own axis and it is arranged to always touch the wall of the stator at the point shown between the inlet tube and the outlet valve. A slot is cut diametrically across the rotor and two vanes are able to slide in the slit. The vanes are separated by springs which always press them against the walls of the stator. Thus, the space between the walls of the stator and the rotor is divided into two separate compartments, labelled A and B.

If the rotor is revolving in an anti-clockwise direction as shown, the volume of A increases and air flows in through the inlet tube from the system being evacuated. At the same time, the air which is trapped in B is compressed and ultimately forced out through the outlet valve. This action is continually repeated as the vanes sweep round and thus the pressure in the system is gradually reduced.

The complete pump is immersed in oil as shown. This not only lubricates the pump but also prevents gas leaking into the high vacuum and helps in cooling the pump. A non-return valve is also incorporated to ensure that air is not sucked from the pump into the evacuated system when the pump is switched off.

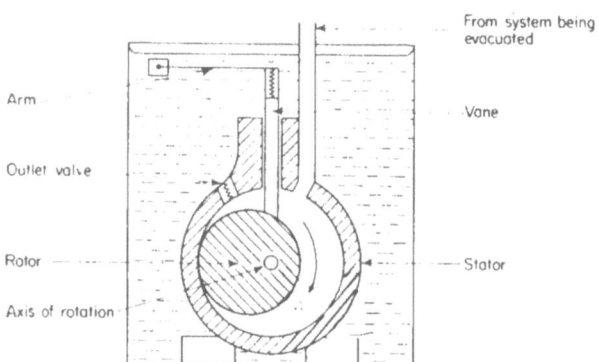

Figure 10.3. Cenco rotary oil pump

An alternative design of mechanical pump is shown in *Figure 10.3.* This is the Cenco-Hyvac pump and it, also, comprises a solid rotor which rotates eccentrically inside the steel stator. A single vane is free to slide through the side of the stator and spring arm always presses it against the rotor. If the rotor is revolving in a clockwise direction as shown, gas from the system being evacuated first fills the compartment A *via* the inlet tube. As the rotor passes the inlet tube the gas in the volume A is trapped and further rotation of the rotor causes it to be compressed until finally it is forced out through the outlet valve. It is common practice to use two Cenco pumps in series, both units being driven by a common shaft. The first pump operates from atmospheric pressure while the second

260

uses the vacuum produced by the first as its backing pressure. This arrangement enables one to attain a pressure of about 10^{-3} torr.

10.2.2 Molecular pumps

The principle of the molecular pump is based on the fact that when the mean free path of gas molecules is larger than the linear dimensions of the apparatus in which they are contained, the molecules make many more collisions with the walls of the apparatus than with each other and thus acquire the properties of the walls of the apparatus. Hence, if a rapidly rotating surface is very close to a stationary one so that the clearance between them is less than the mean free path of the gas molecules contained between the surfaces, the molecules gradually acquire the drift velocity of the rotating surface. If the distance between two parallel boundary planes is t, one of the planes being fixed while the other moves with constant velocity, then, from kinetic theory, it can be shown that for a gas at low pressure, the ratio of the pressure at two points, distant d apart in the direction of motion is

$$\frac{P_1}{P_2} = \frac{cd}{t} \qquad \qquad \ldots (10.6)$$

where c is a constant. It is evident from equation (10.6) that for any given velocity of the moving surface the ratio of the two pressures is constant for a given set of parameters. To attain a low pressure

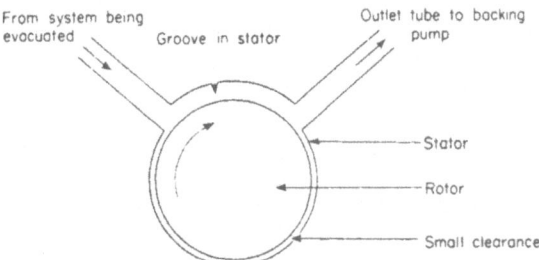

Figure 10.4. Molecular pump

with a molecular pump, therefore, it must be operated in conjunction with a backing pump so that the fore-vacuum in the molecular pump is as small as possible.

Figure 10.4 shows the principle of operation of the pump. The solid cylindrical rotor rotates at a high velocity within the cylindrical steel stator as shown. The gas molecules entering from the system being evacuated acquire a velocity due to impact with the rotor and are swept along the groove in the stator to be finally expelled through the outlet pipe to the backing pump. The rotor speed should be very high and in early pumps it was a few thousand revolutions per minute.

The idea of the molecular pump was first suggested by Gaede, and in 1912 he designed a working molecular pump. This had a set of projections, from the stator, which fitted into grooves in the rotor. Subsequent designs of pump decreased the gap between the rotor and stator to about 2×10^{-2} mm and worked at an angular velocity of 10,000 rev min^{-1}. With a backing pressure of one or two torr a pressure of 10^{-6} torr could be reached. However, the small clearance between the rotor and stator gave rise to mechanical trouble. Modern designs of molecular pump use a much larger clearance, of the order of 1 mm, thus avoiding the mechanical trouble formerly experienced. They also work at higher angular velocities and a modern turbo-molecular pump can work at speeds up up to 16,000 rev min^{-1} With a fore-vacuum of 10^{-2} torr these modern pumps make it possible for a pressure of better than 10^{-9} torr to be reached with air. They are of particular value when it is essential to have a completely oil free vacuum.

10.2.3 Diffusion–condensation pumps

In 1915 Gaede suggested the idea of the diffusion pump for producing low pressures and, as a result of further development, the diffusion–condensation pump is now the most commonly used pump for producing a high vacuum.

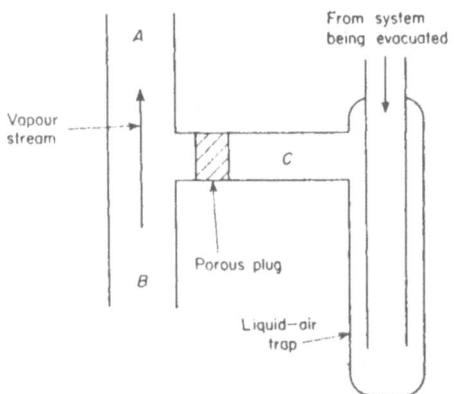

Figure 10.5. Principle of diffusion–condensation pump

The principle of the diffusion pump is illustrated in *Figure 10.5*. If a stream of vapour flows along the tube *AB*, as shown, it carries with it all the gas in the tube. The gas pressure in the tube is therefore reduced and gas diffuses into *AB* from the side tube *C*. As the gas molecules enter *AB* they are also swept along by the vapour stream. Any vapour which diffuses through the plug into the tube *C* is condensed by the liquid-air trap shown. Theoretically, if the vapour stream is free of gas, the gas from *C* will continuously diffuse through the plug until, finally, the gas pressure in *C* is zero. While this is never attained in practice, very low pressures can be reached with the diffusion–condensation pump.

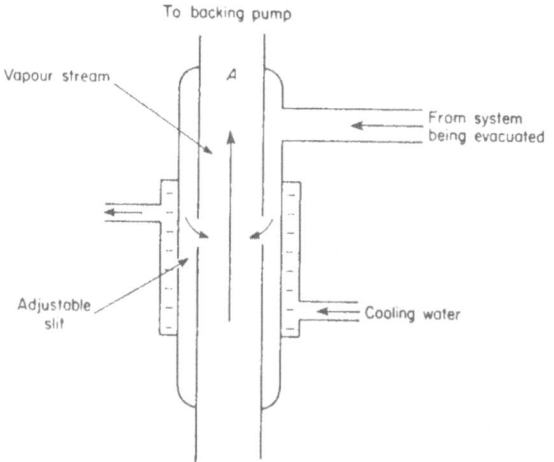

Figure 10.6. Gaede's diffusion pump

The first diffusion pump based on this principle was designed by Gaede. It is illustrated in *Figure 10.6*. Gaede replaced the porous plug by an adjustable slit and cooled the tube surrounding the slit, with water. A stream of mercury vapour passes along the tube *AB* and gas molecules passing through the slit are swept along with the vapour. Ultimately the vapour is condensed and the gas passes to the backing pump. Gaede adjusted the width of the slit in order to reduce back-diffusion of the mercury vapour and showed that for maximum efficiency of operation of the pump, the slit width must be of the same order of size as the mean free path of the molecules in the back-streaming vapour. The water cooling also assists in preventing the vapour from getting through to the system being

263

evacuated. To produce a pressure lower than the vapour pressure of mercury at room temperature, it is necessary to utilize a liquid-air trap between the pump and the system being evacuated. This also has the effect of reducing the pumping speed. The disadvantages of this diffusion pump are the slow pumping speed and the careful control of the mercury vapour temperature which is necessary.

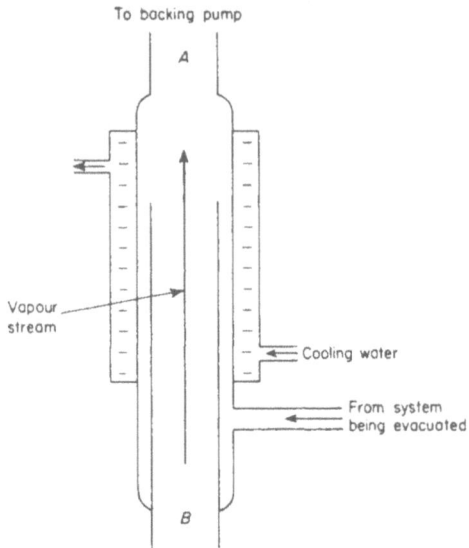

Figure 10.7. Langmuir's diffusion–condensation pump

These disadvantages were both overcome in a modification of Gaede's diffusion pump by Langmuir in 1916. Langmuir directed the mercury vapour stream in a high speed jet as far away as possible from the gas inlet and also, by means of a cooling arrangement, condensed the mercury vapour when it reached the walls of the tube thus preventing it from diffusing through to the system being evacuated. The principle of Langmuir's diffusion–condensation pump is shown in *Figure 10.7*. If the vapour pressure at the annular gap is high then some of the vapour molecules diffuse towards the gas inlet opposing the gas diffusion. However, by keeping a suffi- ciently low pressure at *A*, by means of the backing pump, it is possible to keep the vapour pressure at the gap low enough to prevent back diffusion of the vapour. If the backing pressure is not very high then the speed of the vapour stream must be increased and the annular gap-width reduced.

The first diffusion–condensation pumps were made from glass but modern mercury vapour pumps are made of metal. *Figure 10.8* illustrates the Kaye diffusion pump made from steel. Mercury vapour is produced in the boiler and passes up the central tube to be deflected by the cowl device at the top. A jet of vapour then issues from the annular gap and gas molecules, diffusing downwards

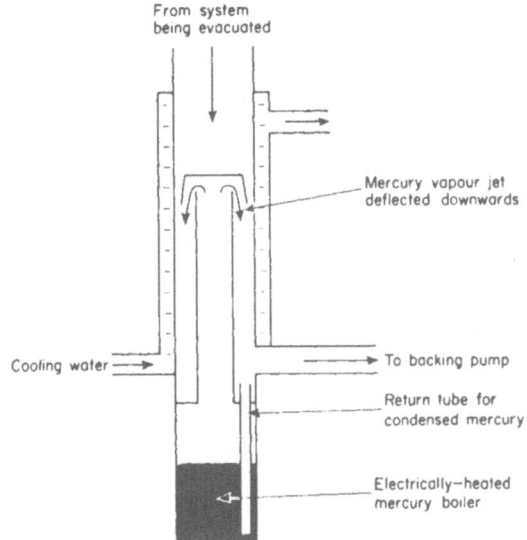

From system
being evacuated

Mercury vapour jet
deflected downwards

Cooling water

To backing pump

Return tube for
condensed mercury

Electrically–heated
mercury boiler

Figure 10.8. Kaye diffusion pump

from the system being evacuated, are swept along by the vapour stream towards the backing pump. The gas-vapour mixture is cooled so that the mercury is condensed and returns to the boiler while the gas molecules are carried away to the backing pump. The Kaye pump has a high pumping speed and will produce a pressure of 10^{-6} torr. Other pumps of this type utilize several gaps and jets of mercury vapour working in parallel.

10.2.4 Oil diffusion pumps

While mercury is a suitable material to use in a diffusion pump on account of its chemical stability, due to the fact that its vapour

265

pressure at room temperature is about 10^{-3} mm, a liquid air trap must be used in conjunction with the pump in order to attain a final pressure in the system lower than this. The necessity for using a liquid air trap was overcome in 1928 when Burch suggested replacing mercury with a high boiling point oil. The oil used has a lower vapour pressure than mercury, of the order of 10^{-6} mm at room temperature, and hence the liquid air trap may be dispensed with, thereby enabling the pump to work at its maximum speed.

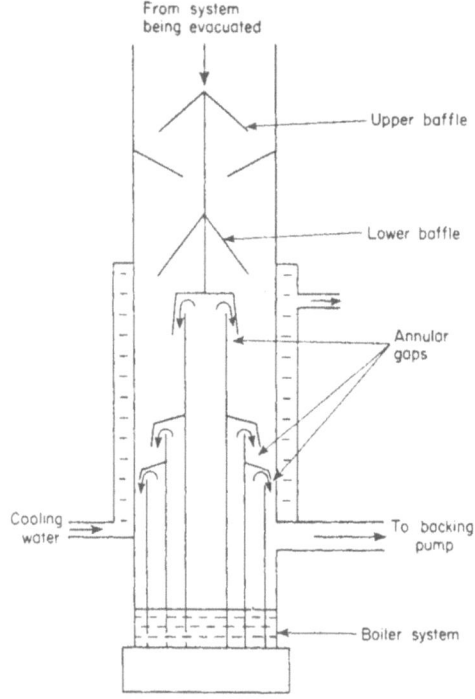

Figure 10.9. Oil diffusion pump

The development of oil diffusion pumps has proceeded steadily since 1928 and today there are several designs of pump available. Generally they use an Apiezon oil or an organic compound such as butyl pthalate. A typical oil diffusion pump is shown in *Figure 10.9*. It will be seen that several annular gaps and jets are used as well as two sets of baffles, the purpose of these baffles being to prevent the oil molecules from diffusing into the system being evacuated. The boiler system is specially designed to provide degassing and fractionation of the oil before it returns to the boiler. This degassing and fractionation is necessary in order to remove, from the oil, volatile impurities which would seriously impair the efficiency of the pump.

The disadvantages of oil diffusion pumps are mainly due to the fact that oils are not as stable as mercury so that if the oil is overheated, due to inadequate cooling or other causes, it tends to decompose. Also oil cannot be heated as much as mercury which means that the pressure in the oil-vapour stream is less than in a mercury vapour stream and hence the backing pressure must be correspondingly less. On the other hand, in addition to the advantage of not requiring a liquid-air trap, greater pumping speeds are possible than with mercury vapour pumps of a similar size, since the molecular weight of the oil is several hundred times greater than that of mercury.

10.2.5 Getter-ion pumps

Getter-ion pumps are based on the ability of certain metals such as barium, titanium and zirconium, provided they are in a finely dispersed form, to absorb a considerable amount of gas. This absorption takes place at room temperature and at other temperatures, and may occur in the following ways:

(1) By a chemical reaction between the gas and the metal particles —'chemisorption'.

(2) By the trapping of the gas on the surface—'adsorption'.

(3) By the occlusion of the gas molecules in the porous layer—'absorption'.

(4) In the case of noble gases, by accelerating the ions, at high energy, towards the gettering wall where they are absorbed as neutralized atoms.

All these processes normally take place at the same time and are generally referred to by the term 'gettering'.

Getter-ion pumps are suitable for applications where a vacuum completely free of oil is required and where fairly small quantities of gas have to be removed over long periods. Getter-ion pumps operate more efficiently as the fore-vacuum is reduced and normally the fore-vacuum must be of the order of 10^{-4} torr. Two different types of pump are available—getter-ion evaporator pumps and getter-ion sputtering pumps.

In evaporator pumps the getter material is thermally evaporated to be deposited on a cooled glass or metal wall. The pump is also provided with an ionization system in order to ionize the gas to be

267

removed. As the gas is absorbed by the getter material according to the processes outlined above, the getter material ultimately becomes exhausted. Hence the pumping speed is not constant over a period of time. The getter material is usually renewed by intermittent evaporation as required so that the getter material layer is constantly built up. In modern pumps the evaporation is controlled by the pressure in the pump, in order to automatically provide the correct amount of getter material.

In getter-ion sputtering pumps there is no thermal evaporation system. The getter film is produced by sputtering a thick titanium plate cathode by means of a gas discharge. Sputtered titanium is deposited on the anode and other parts of the pump thus forming an absorbing film. The formation of the getter material layer takes place concurrently with ionization of the gas. The discharge current is proportional to the gas pressure in the pump and thus the sputtering pump controls the rate of usage of the getter material according to the working conditions.

Getter-ion pumps can be directly connected to the system being evacuated without intervening baffles and liquid air traps. However, it is usual to provide a simple baffle in order to prevent any getter material vapour from entering the system.

10.3 THE MEASUREMENT OF LOW PRESSURES

There are many different types of vacuum gauge available today, enabling one to measure pressures as low as 10^{-13} torr. It must, of course, be realized that to cover this range, several different gauges are necessary and that no single vacuum gauge can be used quantitatively over the entire range. Only two gauges provide absolute measurement of the gas pressure and the other gauges must be calibrated against these absolute instruments.

10.3.1 The McLeod gauge

The McLeod gauge is the standard instrument against which nearly all other low pressure gauges are calibrated. It was developed in 1874 and it is based on the principle of taking a known volume of gas at the unknown pressure, isolating it, and compressing it into a small volume whose pressure is then measured. It consists of a sealed graduated capillary tube attached to a bulb, the volume of which is determined by the range of pressures to be measured. A side capillary tube of similar dimensions is placed next to the capillary tube, as shown in *Figure 10.10*. A reservoir of mercury is connected to the bulb and side capillary and it is thereby possible to trap the volume of gas contained in the bulb and attached capillary by raising the mercury level to be above the base of the bulb, A, as shown. If the volume of the bulb and capillary is V and the gas contained therein is initially at a pressure P, then if it is compressed to a measurable volume v in the capillary tube at a pressure p

$$PV = pv \qquad \qquad \dots (10.6)$$

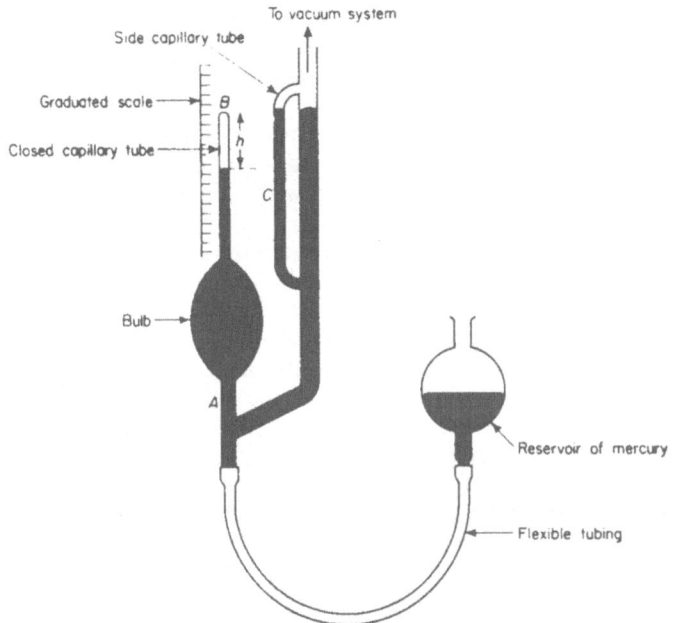

Figure 10.10. McLeod gauge

The McLeod gauge is connected to the system wherein the pressure is to be measured and the mercury reservoir adjusted until the mercury level in the side capillary tube C is at the same horizontal position as the closed end of the capillary tube B. If the difference in heights of the mercury levels in the two capillary tubes is h, this is also equal to the length of the column of gas trapped in the capillary. If the volume of a 1 mm length of the capillary tube is k, then a length h corresponds to a volume kh. Thus, if the volume of the bulb and capillary is V, the required pressure P is given by

$$P = \frac{kh^2}{V} = Kh^2 \qquad \qquad \dots (10.7)$$

269

since $v = kh$ and $p = h$. Provided the gauge constant K is known, P can be measured direct in millimetres of mercury. The use of a side capillary tube, of a similar diameter to the closed capillary, overcomes the error which would arise due to the difference in capillary depression of mercury in two tubes of different diameters.

The McLeod gauge is simple to use and gives an absolute measurement of the pressure. For accurate work, however, several refinements of the simple gauge described above are necessary, and various precautions must be taken. It is normal practice to replace the flexible rubber tubing connecting the bulb and the reservoir with a stainless steel tube dipping into a stainless steel reservoir. This avoids the possibility of air diffusing through the rubber and ultimately getting into the gauge. In addition, over a period of time, the mercury becomes contaminated with sulphur from the rubber tubing. To prevent mercury vapour from entering the vacuum system, a liquid-air trap is placed between the gauge and the system. The gauge is based on the validity of Boyle's law and hence it is only accurate when used to measure the pressure of permanent gases. If condensible vapours are present the gauge reading is not reliable. A further disadvantage is the absence of a continuous record of the pressure, the gauge having to be adjusted and the pressure read at intervals. If all precautions are taken it is possible to measure accurately pressures of the order of 10^{-4} torr to 10^{-5} torr with the McLeod gauge and, as mentioned previously, its main use is in the calibration of other gauges which are more convenient for everyday use.

10.3.2 Knudsen's absolute pressure gauge

The Knudsen gauge provides an alternative means of determining absolutely the pressure of a gas. It has an advantage over the McLeod gauge in that it can also be used for measuring the pressure of vapours. It has the disadvantage, however, of being much more difficult to use.

The instrument is illustrated in *Figure 10.11* and consists of two fixed mica strips, A and B, heated electrically and placed on either side of a third strip C, suspended by a quartz fibre from a torsion head. A mirror is attached to the fibre and the deflection of the suspended strip found in the usual manner. The gauge is contained in a glass container which connects with the vacuum system whose pressure is to be measured.

The theory of the gauge is as follows. Let the temperature of the fixed mica strips be T_1 and that of the suspended strip be T_2. Let the number of gas molecules per unit volume, moving from A to C

with root mean square velocity c_1, be n_1 and the number per unit volume moving from C to A with root mean square velocity c_2, be n_2. Let the containing vessel be at temperature T_2 and let the number of molecules per unit volume in the space other than that between A and C be n. Then, when equilibrium is reached, the number of molecules impinging on unit area in unit time must be the same on both A and C, so that

$$n_1 c_1 = n_2 c_2 \qquad \qquad \dots (10.8)$$

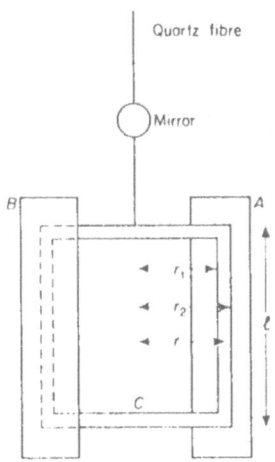

Figure 10.11. Knudsen pressure gauge

Also, the number of molecules leaving the space between A and C is equal to the number flowing in, hence

$$nc_2 = n_1 c_1 + n_2 c_2 \qquad \dots (10.9)$$

From kinetic theory, the pressure acting between A and C is given by

$$\tfrac{1}{3}mn_1 c_1^2 + \tfrac{1}{3}mn_2 c_2^2 \qquad \dots (10.10)$$

where m is the mass of a gas molecule. Also, the pressure acting on the face of C on the side removed from A is $\tfrac{1}{3}mnc_2^2$, hence, the excess pressure acting on C, repelling it from A, is

$$\tfrac{1}{3}mn_1 c_1^2 + \tfrac{1}{3}mn_2 c_2^2 - \tfrac{1}{3}mnc_2^2 \qquad \dots (10.11)$$

Substituting for n_1 and n_2 from equation (10.9)

271

$$F = \frac{1}{2} P \left(\frac{c_1}{c_2} - 1 \right) \qquad \qquad \dots (10.12)$$

where $P = \frac{1}{3} mnc_2^2$ and is the pressure in the container. F is thus equal to the force per unit area acting on C, repelling it from A. If it is further assumed that molecules striking A and C assume the temperature of these surfaces, then

$$\frac{c_1}{c_2} = \sqrt{\frac{T_1}{T_2}}$$

so that
$$F = \frac{P}{2} \left(\sqrt{\frac{T_1}{T_2}} - 1 \right) \qquad \qquad \dots (10.13)$$

If the temperature difference is not greater than about 250 degC this can be approximated to

$$F = \frac{P}{4} \left(\frac{T_1 - T_2}{T_2} \right) \qquad \qquad \dots (10.14)$$

If the distances of the vertical sides of C from the axis of suspension are r_1 and r_2, the length of the vertical side is l, the angle of deflection of C is θ and the torsional constant of the suspension fibre is τ, then the total force due to both strips causing C to twist is $2 \int_{r_1}^{r_2} Flr\, dr$ where dr is the width of a small vertical strip, and hence

$$\tau\theta = 2 \int_{r_1}^{r_2} Flr\, dr = Fl(r_2^2 - r_1^2) \qquad \qquad \dots (10.15)$$

Hence
$$\tau\theta = 2AFr$$

and
$$F = \tau\theta/2Ar \qquad \qquad \dots (10.16)$$

where A is the area of one vertical strip and r is the average distance from the axis of suspension. Substituting for F in equation (10.14) gives

$$P = \frac{2\tau\theta}{Ar} \cdot \frac{T_2}{T_1 - T_2} \qquad \qquad \dots (10.17)$$

If the moment of inertia of the suspension system is I, and the periodic time of oscillation is t, then

$$t = 2\pi \sqrt{\frac{I}{\tau}}$$

so that finally

$$P = \frac{8\pi^2 I\theta}{Art^2} \cdot \frac{T_2}{T_1 - T_2} \qquad \qquad \dots (10.18)$$

272

The Knudsen gauge is useful for measuring pressures below 10^{-2} torr. It begins to behave erratically when the gas pressure becomes so high as to cause convection currents to be set up.

As well as being usable with condensible vapours, the Knudsen gauge has many other advantages over the McLeod gauge as an absolute instrument. It is more sensitive than the McLeod gauge at low pressures and it is also extremely stable. No mercury vapour is involved, thereby eliminating the need for a liquid-air trap between the gauge and the vacuum system. It also provides a continuous reading of the pressure in the system. However, in use it is not so convenient as other types of secondary gauge described below and consequently it is only used for calibration purposes.

10.3.3 The Pirani gauge

While at normal pressures the thermal conductivity of a gas is independent of the gas pressure, it can be shown from kinetic theory that at low pressures the thermal conductivity of a gas is directly proportional to the gas pressure, so that

$$k = \gamma P \qquad \ldots (10.19)$$

where k is the thermal conductivity of the gas, P is the gas pressure and γ is a constant. This relationship is made use of in the Pirani gauge, which is used for measuring the pressure of gases below 10^{-2} torr.

The gas pressure is determined indirectly by measuring the change of electrical resistance of a heated wire placed in the gas. Usually a fine tungsten wire is used with a high temperature-resistance coefficient. The wire is connected in one arm of a Wheatstone bridge circuit and a constant potential difference, sufficient to heat the wire to several degrees above its surroundings, is applied to the bridge. The rate of heat transfer from the wire to its surroundings determines the temperature of the wire and hence its electrical resistance. Thus, the electrical resistance is proportional to the gas pressure and once the gauge has been calibrated by means of a McLeod gauge, the gas pressure can be determined from the resistance measurement, i.e. if the potential difference across the filament is kept constant and the current flowing is observed, then the resistance can be calculated. As the gas pressure decreases, the resistance of the wire increases and hence the current flowing decreases. Actually, the calibration of the gauge is only valid if the conditions under which the calibration was carried out are maintained. To comply with this requirement, Campbell modified the original Pirani gauge by keeping the temperature of

273

the wire, and hence its resistance, constant. Thus, instead of measuring the variation of current flowing through the wire, Campbell measured the change in the potential difference applied to the wire in order to keep the temperature constant.

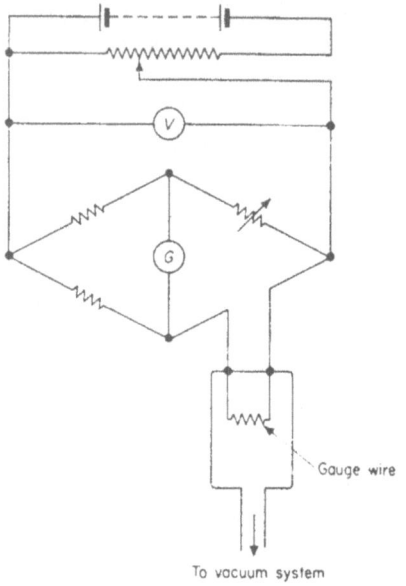

To vacuum system

Figure 10.12. Pirani gauge

Figure 10.12 shows the circuit used in this arrangement. The gauge wire is maintained at a temperature of about 100°C and is placed in one arm of the bridge. In the other three arms are manganin resistances of practically zero temperature coefficient. A voltmeter is connected across the bridge and the potential difference across the bridge is adjusted by the rheostat in the battery circuit until balance is obtained. If the potential difference across the bridge is V, the rate of heat dissipation in the tungsten wire is αV^2, where α is a constant. If the rate of loss of heat by conduction through the gas is represented by some function of the gas pressure, $f(P)$, then

$$\alpha V^2 = \beta T + f(P) \qquad \qquad \dots (10.20)$$

where T is the excess temperature of the wire over its surroundings and βT represents the heat losses along the leads to the wire, β

274

being a constant. If the potential difference across the bridge is V_0 when the pressure is zero, then

$$\alpha V_0^2 = \beta T \qquad \qquad \ldots (10.21)$$

Combining equations (10.20) and (10.21) gives

$$\frac{V^2 - V_0^2}{V_0^2} = \frac{1}{\beta T} f(P) = C f(P) \qquad \ldots (10.22)$$

where C is a constant, since $1/\beta T$ is constant over a considerable pressure range. It is found that $f(P)$ is approximately proportional to the pressure P, so that $(V^2 - V_0^2)/V_0^2$ is directly proportional to the gas pressure.

The sensitivity of the Pirani gauge is proportional to the specific heat of the gas and is inversely proportional to the square root of the molecular weight of the gas. For accurate work the gauge must be calibrated using the gas it is subsequently going to be used with. For many gases, the calibration curves lie fairly close together but it can be seen that if the gauge is calibrated with air, for example, it will give an incorrect reading of the absolute pressure if, in its subsequent use, any organic vapours are present. The Pirani gauge is usually used for measuring pressures in the range 10^{-2} torr to 10^{-4} torr but with special arrangements pressures down to 10^{-5} torr to 10^{-6} torr can be measured. However the gauge is not often used over this latter range.

10.3.4 The ionization gauge

The ionization gauge is based on the linear dependence of ionization taking place in a gas with the gas pressure. If a small amount of gas is present in a triode valve so that the activity of the filament is unaffected by its presence, then, provided the potential difference between the anode and cathode exceeds the ionization potential of the gas, ionization takes place. Ionization refers to the process whereby a fast electron, liberated from the filament and travelling towards the anode, strikes an atom of the gas and ejects an electron from an outer shell, thereby producing a positive ion of the gas. If the positive ions formed in this manner are collected on a third electrode, i.e. the grid of the valve, then, for a fixed value of the potential difference across the valve, the positive ion current varies with the gas pressure. At a low gas pressure, where the probability of an electron making more than one collision in travelling from the filament to the anode is remote, then the positive ion current varies linearly with the gas pressure.

Theoretically any triode valve can be used as an ionization gauge but usually a special type of construction is used (*Figure 10.13*). The filament is surrounded by a cylindrical grid whose axis is coaxial with the filament axis, while the anode consists of a silver

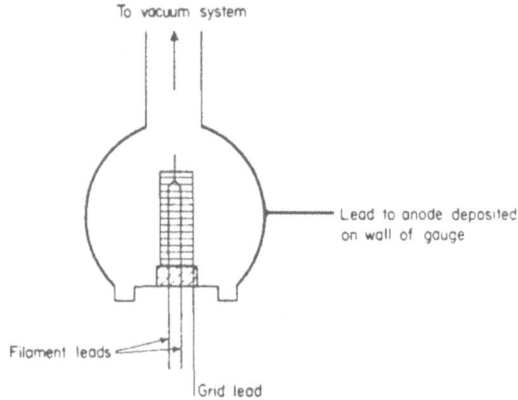

Figure 10.13. Ionization gauge

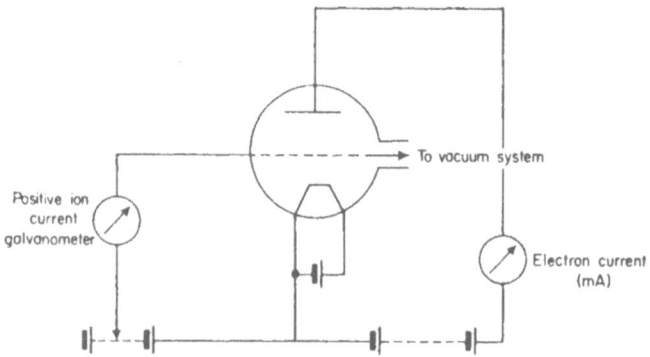

Figure 10.14. Ionization gauge circuit

deposit on the walls of the valve. *Figure 10.14* shows the circuit used with the ionization gauge.

A small negative potential is applied to the grid and the grid current, due to the positive ions flowing to the grid, is measured by the galvanometer shown. The electron current is measured in the anode circuit. The number of positive ions produced in the valve is proportional to the electron current as well as to the gas pressure, and the equation relating these quantities is

$$P = \frac{kI}{i} \qquad \text{.... (10.23)}$$

where, P is the gas pressure, I is the positive ion current, i is the electron current and k is the gauge constant which must be determined by initially calibrating the gauge with a McLeod or Knudsen gauge. Equation (10.23) is found to be valid for pressures below 10^{-3} torr, provided that saturation of the ionization current is not reached.

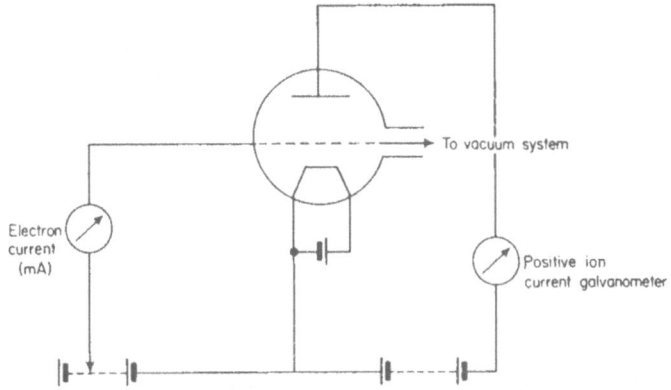

To vacuum system

Electron current (mA)

Positive ion current galvanometer

Figure 10.15. Alternative ionization gauge circuit

Figure 10.15 shows an alternative arrangement which is more sensitive than the basic scheme described above. In this circuit, the grid is kept at a positive potential with respect to the filament and the anode is made negative. Hence, the electron current is measured in the grid circuit while the positive ions are collected at the anode.

10.3.5 Penning vacuum gauges

There are various designs of the Penning vacuum gauge but they all have the common feature of two cold electrodes, anode and cathode, between which a so-called cold discharge is maintained by application of a d.c. voltage of the order of a few kilovolts. The discharge current is a measure of the gas pressure in the gauge head. At low pressures, the discharge current must be maintained by means of an external permanent magnet. The primary electrons which induce the current are emitted as a result of field emission at the cathode. The permanent magnet is arranged so that the magnetic field is perpendicular to the plane of the anode and cathode (*Figure 10.16*). Thus the electrons move in a helical path from the cathode to the anode. This increase in the electron path considerably increases the probability of the gas molecules being ionized and hence a measurable current results, even at very low pressures.

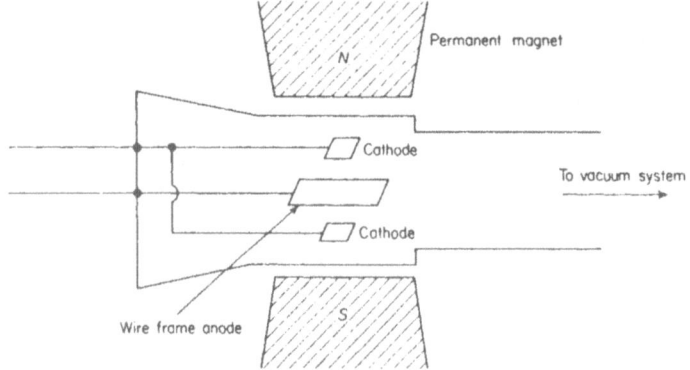

Figure 10.16. Penning gauge

Penning gauges are used for measuring pressures less than 10^{-3} torr. With a suitably constructed gauge, pressures as low as 10^{-13} torr are measurable. The advantage of Penning gauges lies mainly in the absence of a hot cathode, so that they are unaffected by air leaks up to atmospheric pressure. They are also robust and easy to operate and consequently are widely used in industry. A disadvantage is their lack of accuracy. Since the discharge current is initiated by field emission, which is dependent quantitatively on the surface condition of the cathode, the accuracy of Penning gauges is not much better than $\pm 20\%$.

10.3.6 Partial pressure gauges

It is sometimes necessary to know the composition of the gases present in a vacuum system as well as the total pressure exerted. The total pressure is due, of course, to the partial pressures exerted by each component present. This necessity has led to the development of partial pressure gauges such as the Omegatron which allows qualitative and quantitative determinations of partial pressures to be made.

THE MEASUREMENT OF LOW PRESSURES

A partial pressure gauge is really a small but sensitive mass spectrometer. The gauge consists of an ion source and a separating system for the different gas ions produced. The gas ions of different mass are usually separated by utilizing resonance phenomena, with the assistance of magnetic and electric fields. *Figure 10.17* shows the basic construction of the Omegatron partial pressure gauge. The gauge is connected to the vacuum system and any gas present is ionized by an electron beam from the cathode, the diameter of the electron beam being limited by a slit system. The electron beam travels in the direction of the magnetic field which focuses it. The ions produced travel in a spiral path, due to the effect of the magnetic field and the high frequency electric field which is applied at right

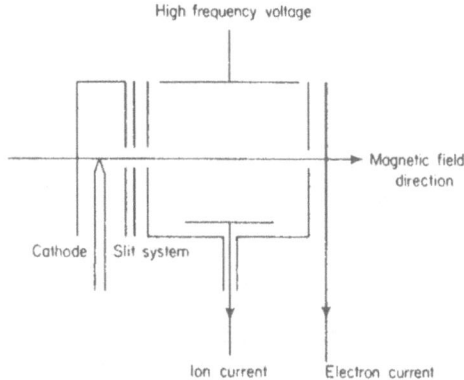

Figure 10.17. Omegatron gauge

angles to the magnetic field. The only ions which remain in phase with the high frequency field are those which satisfy the resonant condition

$$2\pi f = \frac{e}{m_0}\ \frac{B}{M} \qquad \qquad \dots (10.24)$$

where, f is the frequency of the applied field, e/m_0 is the specific charge of a particle having $M = 1$, B is the magnetic induction and M is the mass number in atomic weight units.

The ions which remain in phase with the field continuously absorb energy and spiral along a circular path until they reach the ion collector. The ion current produced is amplified and monitored on an external recorder connected to the amplifier. The spectrum of ions of different mass is recorded by steadily varying the frequency of the applied field. With the Omegatron, when the magnetic field strength is about 0·4 tesla the frequency for hydrogen is about 3 MHz.

An Omegatron gauge has a high sensitivity so that it is very suitable for measuring partial pressures at a very low total pressure. It can be used at a total pressure of 10^{-5} torr and it will measure partial pressures as low as 10^{-11} torr.

There are other partial pressure gauges such as the Topatron, which operates without the presence of an external magnetic field, but a full discussion of these gauges is beyond the scope of this book.

10.4 THE DETECTION OF LEAKS IN A VACUUM SYSTEM

In the production of low pressures, to maintain a high vacuum, i.e. 10^{-6} torr, continuous pumping is necessary. Adsorbed gases are continuously released from the walls of the vacuum system and consequently, if pumping is stopped, the pressure in the system gradually rises. In the case of radio valves and similar devices which are sealed off, it must be ensured that the pressure in the device never rises above an acceptable value. This is normally done by thoroughly degassing the components in the valve and by the incorporation of a getter material.

The release of adsorbed gases from the walls of a vacuum system gradually decreases with time and can easily be dealt with by the vacuum pumps. A more serious obstacle to the attainment of a high vacuum is the presence of leaks in the system. These allow air to flow into the system at a constant rate, or worse, if the leak gets bigger, 'at an increasing rate. Consequently, it is essential to ensure that the system is entirely leakproof or at least to make sure that any leaks are so small that the inflow of air can be dealt with by the pumps, thus maintaining the required pressure in the system.

When a vacuum system is constructed it invariably has some leaks and these must first of all be found and then repaired. On the premise that prevention is better than cure, where possible, individual components of a vacuum system should be tested for leaks before they are connected together. Even so, when the system is assembled, leaks are frequently found. These may be due to various causes, such as seals or gaskets not being properly seated, or components being slightly deformed dusing assembly, etc.

For leak-testing glass vacuum systems, the most usual technique utilizes the high frequency vacuum tester. The technique is to apply a test liquid to the suspected part of the system and then bring the high frequency vacuum tester near. If there is a leak the colour of the discharge changes from pink to pale blue. If the high frequency electrode is brought near to a small pinhole in the glass, then a brightly coloured discharge path will be seen close to the hole. However, it is possible that the pinhole may be enlarged by the discharge and also, if the strength of the electric field is high, thin parts of the system may be fractured. Hence, considerable care is necessary when using the high frequency leak detector.

In most permanent vacuum systems, low pressure gauges, usually ionization gauges, are incorporated into the system at several points as aids to leak detection, as well as for pressure measurement. If a leak is suspected the gauge nearest the suspected part of the system is switched on. The outer surface of the system at the suspected point is sprayed with a heavy test gas, such as acetone, to which the gauge is especially sensitive. Entrance of the gas into the system is rapidly registered by the gauge reading.

More sophisticated leak detection devices have now been developed. These are complicated instruments and are based on mass spectroscopic gas analysis. A full description of such devices may be found in specialist books on high vacuum techniques.

GENERAL INDEX

Ablett,
  determination of angle of contact by
    method of, 159
Acceleration,
  angular,
    constant speed, at, 45
  due to gravity,
    effect of terrestrial latitude on, 72
    measurement of, 54
  effect of gravity on, 70
  particles, of, 1
    unilinear, 2
  radial,
    due to Earth rotation,
      effect of latitude on, 71
  uniform,
    constant speed, at, 45
Adam and Jessop,
  determination of angle of contact by
    method of, 158
Adam's surface tension balance, 187
Air,
  viscosity of,
    determination of, 208
    effect on oscillation of pendulum,
      56, 65
Airy,
  calculation of mass of Earth, by, 93
Amagat's method,
  determination of bulk modulus of
    solids by, 146
Attraction, force of,
  flat plate and liquid surface, between,
    162
  parallel plates separated by thin liquid
    layer, between, 161
Avogadro's number, 252, 254, 256
Axes,
  parallel,
    theorem of, 30
  perpendicular,
    theorem of, 30
Axial modulus, 109

Balance method,
  measurement of surface tension by,
    170
Balances, torsion, 77
  measurement of surface tension by,
    171
Bar, rectangular,
  moment of inertia of,
    central axis, about, 36
Beam,
  bending of, 114, 116, 120
    energy in, 115, 127
  end-supported, centrally loaded, 119
  glass,
    Young's modulus and Poisson's
      ratio of, 135
  material of,
    Young's modulus of, 124, 125
  vertical,
    thrust stress, under, 120
Bending, beam, of, 114, 116, 120
  energy in, 115, 127
Berkeley and Hartley,
  osmotic pressure, method of measure-
    ment of, 239
Bernoulli's theorem, 227
  illustration of, 229
Bingham plastic flow, 219
Body,
  mass of,
    effect on gravity, 83
  response to external forces, 105
  rigid,
    angular momentum of, 39
    example of, 42
    motion of, 25, 29
      example of, 38
    rotating,
      kinetic energy of, 40
Boliden gravity meter, 75
Bouguer,
  calculation of mass of Earth by,
    91

283

Momentum—*cont.*
incompressible matter, of, 225
linear,
change of, 7, 25
rigid body, in, 29
definition of, 6
Morse,
osmotic pressure, method of measurement of, 240
Motion,
Euler's equation of, 225
gyroscopic, 44, 47
particles, of, 25
example of, 4
inverse square law of attraction, under, 13
laws of, 6
rectilinear, 1
two-dimensional, 9
uniformly accelerated unilinear, 2
planets, of,
example of, 18
Kepler's laws of, 17
projectiles, of,
frictionless, 20
subject to resistance, 22
example of, 24
rigid body, of, 25, 29
example of, 38

Neumann's triangle,
surface tension forces, of, 156
Newtonian liquids, 218
Newton's
law of gravitational attraction, 83
law of viscous flow, 190, 218
laws of motion, 6, 25
Nitrogen,
compressibility of, 149
Non-Newtonian liquids, 218
rheological properties of, measurement of, 221

Oil diffusion pumps, 265
Oils, heavy,
viscosity of, determination of, 206
Omegatron gauge, 279
Orbits,
central,
conic, 14
elliptic, parabolic and hyperbolic, 16
particles, of, 11

Orbits—*cont.*
elliptic,
periodic time of, 17
Oscillating disk viscometer, 216
Oscillations,
damped, period of, 56
pendulum, of, 55
effect of magnitude of mass on, 57
vertical,
loaded spring, of, 140
Osmosis, 235
laws of, 236
Osmotic
balance, 241
pressure, 236
measurement of,
Berkeley and Hartley's method of, 239
Morse's method of, 240
solutions, of,
relationship to diffusion, 247
theory of, 241
Ostwald viscometer, 203
Oxygen,
viscosity of, 211

Parallel
axes theorem, 30
plates,
attractive force between, 161
Particles,
Brownian motion of,
Einstein's and Smoluchowski's equation of, 253
gases, in, 254
liquids, in, 249
dynamics of, 1
gravitational potential of, 87
Pendulum,
compound,
description of, 62
oscillation of, 51
example of, 69
determination of gravity-induced acceleration by, 54
corrections required in, 55, 65
examples of, 58
gyrostatic, 48
invariable,
density of Earth calculated by, 93
description of, 73
gravity measurement by, 74